MEDITATIONS FROM THE TEMPLE

A collection of meditations for balance and well-being given by the teachers from THE TEMPLE OF INITIATION INTO THE RADIANT LIFE

Compiled by

GLORIA RUTH RIVERS

Published and distributed by

TEMPLE PUBLICATIONS, INC.
3327 SW Dosch Road
Portland, OR 97201

ISBN: 0-918341-02-7 Limited Edition

Printed in the United States of America

MEDITATIONS FROM THE TEMPLE, a selected compilation of meditations for balance and well-being which teach the art and discipline of using your mind, body and imagination to transform your life. The teachers tell us the goal for humankind is self-mastery and give us these meditations to help us along our path.

This book is for you if you want:

vibrant health, personal confidence, loving relationships, self-control, prosperity and spiritual growth;

enlightenment, out-of-body awareness, greater strength and energy, psychic development, and communication with your own Source Group;

control over your environment, political influence, ultimate world peace, and a clean planet.

These meditations can be used in groups or individually. A designated leader may read the meditations slowly with appropriate pauses, while listeners enjoy the guided meditations and a growth experience. Alone one may read a little, then stop to grasp the import, contemplate the meaning, or otherwise experience the material. Both approaches are valid and stimulating.

TEMPLE PUBLICATIONS, INC.
3327 SW Dosch Road
Portland, OR 97201

ISBN: 0-918341-02-7 Limited Edition

PHOTOGRAPHS
courtesy of Owen Putnam Cramer

selected from his collection of
photographs taken by his mother

HARRIET PARK CRAMER
1887-1940

Most of Mrs. Cramer's work focuses on the Northwest during the 1920's and 1930's. Her photographs are unusual because most of our heritage of photos from this period is of people rather than nature.

Mrs. Cramer exhibited an unusual artistry in the selection of sites for her pictures, and a stunning conceptualization in their framing. We are most privileged to have her work to enhance the experience of the meditations.

In recognition:

LaMar Bell, whose initial concept that a collection of meditations from the transcripts might be helpful to us all;

Owen Cramer, whose encouragement and support made this effort possible

Richard White, Owen Cramer and Sidney Smither who assisted with the initial meditation selections and titles

Sidney Smither, Betty Welty, and Owen Cramer, who did much of the proofreading

The editorial committee, Richard White, Owen Cramer, Sidney Smither, Jean Chisholm, Myra Perala, Jenika Jefferson, who made the initial selection of photographs

Thanks to all of you.

Dear Reader:

This book has something for everyone depending on what one is looking for. The meditations were chosen to be demonstrative of the quality of teaching we are receiving from The Temple, as well as to be of intrinsic value to the user.

This book is truly a heart-crafted product as well as a hand-crafted product. Some of the earliest meditations were transcribed by hand by a circle member from the tapings before they were typed. All of the typesetting for this book I have done myself on my Tandy 1000 with a DWP220 printer. The headings were hand set by me. All the design, layout and paste up can be laid at my door. Therefore, as in all hand-crafted items there are imperfections. In some instances the imperfections were left in by design just so there would indeed be something for everyone. How can one criticize something perfect?

The book is printed on Delo Opaque 60# English Finish paper and the cover is 67# Exact Vellum Bristol. The actual printing was done on a five-year-old A.B.Dick Model 375, Super-Aquamatic, chain delivery. These little items are important to some people! Sometimes, in the wee small hours of the morning, minor adjustments were omitted, so some of the margins don't match, and the page numbers aren't back to back. But whose counting?

All the pages were hand-collated and hand bound in my shop by me and my niece, Ardyth. We also did the shrink wrapping and packaging. Believe me -- for all its imperfections -- you are holding in your hand my blood, sweat and tears -- and my heart. Bless you in Light and Love.

Gloria Ruth Rivers

TABLE OF CONTENTS

PREFACE

Source of the Meditations

Presented by several teachers over a period of four years, these meditations were given in connection with lessons on the spiritual nature of physical existence. The teachers have made it clear that the Temple is not a religious institution, but it is an institution of learning.

Communication with these teachers is by trance channeling. Early in the experience of our circle, one of our teachers, Karansa, told us that all our communicators are members of the Temple of Initiation. As students, we are also numbered in its membership. Later we were told,

"We, of the spirit-physical*, are the Initiates of the Cosmic Soul...Initiates are those who have restored their heritage to each realm through knowledge and understanding...

"The Temple of Initiation is a Temple of Spiritual Consciousness for you and a Temple of Divine Consciousness for us. It is a Temple of Restoration for all of us... So let us, then, call our work 'Temple of Initiation into the Radiant Life'".

The Temple of Initiation faculty is in a realm they call the Spirit-Physical . One of the groups they work with is in the Portland,

---

* See glossary

Oregon area where Gloria Ruth Rivers is the trance medium who provides the channel. That group, calling themselves the Temple Circle, has incorporated as Temple Publications, Inc. for the purpose of making the channeled Temple information available to a wider public.

There have been over one hundred channeling sessions that follow a structured course of study. Sessions end with a guided meditation, frequently related to the material presented. Most meditations include some form of personal healing and building of energy. Some provide direct loving assistance to those involved in traumatic situations or events -- bereavements, hijacking, the shuttle explosion. The loving, healing energy generated is always shared with those for whom Circle members have concern, and is then amplified and released to the world.

The Temple faculty members presenting meditations include:

-- Judith, who has presented many profound lessons on the nature of our existence and the spiritual hierarchy.

-- Silverthistle, who is a scholar of the holy scriptures.

-- Sheariam, who handles most of the general instruction.

-- Joshua, an addition to Gloria Ruth's spirit guides who lends his great energy to the successful reception and promulgation of the Temple teachings.

The work of the Temple is overseen by a council comprised of entities from what the teachers call "the Wise Ones" of whom the White Brotherhood is a part.

Also mentioned in the meditations are 'the Temple healers'. We have not been introduced to individual healers, but often during the meditation period, they come as a group to help us. The teachers have said that when we sit in meditation and ask for healing, the healers join us even though a Temple teacher is not present. We have been assured that this is true for anyone, not exclusively the members of the circle. So we have included those meditations which involve the Temple healers.

References at the end of the meditations and in footnotes (such as S-42) refer to the circle transcript from which the meditation is taken. The teacher's name is also given. These references are included for the benefit of those who have a set of the transcripts and wish to look up the session during which the meditation was given.

Other references such as "Glossary", "Index" or "Appendix" are to be found in the back of the book.

INTRODUCTION

The Temple Teachers
on Meditation

The meditations are presented as received (taped and transcribed) except for an occasional personal reference that has been changed to general usage terms, explanatory material necessary for understanding, and syntax changes needed for clarity.

Sheariam discussed the form of meditation presented here[1], the guided meditation:

"The importance of meditation is that the attention be redirected away from the daily material concerns and toward spiritual growth, the finding of one's Self, and the actualization of the Divine Self. We use the form of meditation as given in these sessions in order to keep the attention of all members present directed in a specific focus. In groups, this is one of the better means to accomplish a specific end."

Then she continued: "As the individuals of a group reach agreement upon a format and agreement upon the desired results, then any kind of meditation decided upon should achieve those results whether it is by guided meditation, or by silent meditation, or by mantra, or by listening to music. Whatever the means, it is in the Realm of Mind[2] that the goal is established. And when minds meet, you achieve great power."

At another time[3] Sheariam said, "Very often the

1. Ref. Session #52
2. See Temple Teachings Book II
3. Ref. Session #50

one who feels unable to meditate does not recognize meditation. There are abroad in your culture today many different definitions of meditation, recommendations of how it must be done, of what it consists, etc. Meditation is merely an action of the mind. It is that action during which there is some perception of knowledge beyond that which had previously occupied the thoughts. You have participated in meditation any time you get an idea.

"The meditation we recommend is one of contemplation. You set aside a quiet time in which you allow yourself to consider some precept. And when you are through with that consideration, there should be greater understanding of the precept because there has been introduction of new information from a higher source.

"As you meditate, as you sit quietly, what you must clear from your minds is your daily concerns -- do not be concerned about unwashed dishes, unmade beds, or unwritten letters. Purpose yourselves to consider something of importance, one of the (Cosmic) laws, for instance. And ponder, consider, contemplate, ask. It need not be directed to a specific individual intelligence. You can make a blanket demand to know. Say to yourself, 'Source, I am knocking. I am asking. I demand to know.' And then allow yourself to be taught by focusing your attention on one thing."

Early in the presentations, Silverthistle emphasized the importance of meditation:

"Your quiet time is essential to your initiation process. It is in such a period of at-rest-ment that your own teachers can make direct communication with you. You people in the flesh tend to allow physical concerns to cheat you out of contemplative time, thus you retard your own progress. You should make a definite effort to spend a few minutes in each day in a receptive state of mind. Do not become discouraged if nothing seems to happen for a lengthy period. It takes time (for your teachers) to enter your vibrations sufficiently strong enough to inspire you with an idea. But it is those teachers who come to you in dreams and hunches and other perceptive devices to give you guidance.

"You have been admonished to seek the wee small voice within. And that is how we as teachers can speak to you directly. Each novitiate is given a certain assigned group of teachers and guides and monitors to urge progress, to teach that which is to be learned at whatever place one is in one's development. The best effort you can offer is the desire to learn and to provide the time in which it can happen."

More recently Joshua also spoke on the importance of meditation:

"But use meditation; use contemplation. Allow yourselves freedom of your thoughts. Do not continue to pollute your mental and emotional atmosphere with the technology available to you today. Allow yourselves time to be away from telephones, from televisions, from radios, from stereos, from the pollutions of society, from the pollutions of aggregate consciousness.....

"There is abroad in your society today a practice of arousing fear and fright as a means for seeking that peak experience, that plutonian experience. The thrill and joy of transcendence through meditation and contemplation can take you into realms which no drug could provide for you, whether it is a chemical drug, a sound drug, or an emotional drug. As you seek the spiritual path, you come to know that spiritual being which you are. It isn't a case of becoming spiritualized; it's a case of remembering who you are."

The Golden Bubble

Unburden your bodies. Put down your accoutrements. Bring physical balance into the physical parts of your vehicle. Breathe deeply. Allow all tension to depart from you.

With your powers of visualization, imagine being encapsulated in a golden bubble. You can see out. You can be seen from without. But you are safe; you are secure within the golden bubble of the Divine Dewdrop. You are nestled in the very bosom of God, where there is perfect balance, perfect peace, perfect understanding.

As you nestle within this golden globe, allow that globe to extend to meet your neighbors'. Like bubbles, the bubbles open and all become one within one huge golden bubble.

In this Divine Breast there is no fear -- only peace and joy. There is no imperfection, only wholeness and repleteness.

The Divine Energy of Life flows through you, flows throughout the bubble.

Let now that bubble expand and expand so that the Divine Idea of Life in Earth moves to encircle the entire planet. Thus the Divine Idea serves all needs everywhere for everyone. Yet your portion is not diminished -- you have as much as you ever had.

As you go through your daily routines, whenever a concern should arise, think again of this golden bubble. Allow yourselves to nestle within the Bosom of God, and know that the Divine Love conquers all.

As you extend this sense of peace and wholeness around the world, those about whom you have concern are served -- served properly for their need, as only the Divine knows the within of them.

So you are blessed. Take this blessing with you always. It dwells with you, and it is in you. You share it by being yourself.

So I leave you now with my blessing, knowing that we are not separated. We are always one in the Divine Golden Bubble. Bless you all.
(Ref. S-40 Silverthistle)

Cleansing, Energizing, Healing

Work with the ball of energy* between your palms until you have a good strong pulsation. Extend your hands before you, placing them an inch apart. Then slowly rotate them until you begin to feel the energy build.

As you do this, allow your bodies to relax. Breathe deeply, regularly. As the energy builds between the palms of your hands, allow your hands to move farther apart, until there is about five or six inches between them.

Then as the energy builds, let your exhalation blow gently upon that ball of energy. Know that the very vital essence of your own personal energy, the love that you hold for your fellowman, is going into that ball of energy with the very breath of your body. You inhale God's breath. You exhale God's love.

Open your eyes and look at the ball of energy. You should be able to see it to some extent. It should be sparkly greens, blues and roses. Enjoy the color. Just sort of roll the ball back and forth in your palms and enjoy the sparkling evanescent color.

Now gently, gently... blow the ball of energy like a bubble so that it joins all the other balls of energy in the middle of the room. Just open your hands; gently blow the ball of energy.

---

*See Appendix for "Energy"

Then let your hands rest, palms upward on your laps. Just regard that ball of energy... evanescent, moving, powerful. Just think of all the healing that rests within that sphere.

Now, each one partake of that ball of energy by extending your toes out toward it. Draw the energy into your bodies through your toes. Feel it as it comes into your feet and rises up your legs. Your legs tingle, becoming warm. The energy rises up to your abdomen and you feel a thrill of joy. There is an excitement that builds within you as the energy rises up.

As it moves through your body, it cleanses all of the tissues; any organ that is out of order is energized and brought back into balance. Where there is excessive fluid in the tissues, the energy absorbs it and the body is free.

Where an organ is weak in energy, the healing energy deposits a residue -- builds up the battery of the body.

Then the energy rises up through the chest cavity, up through the throat and into the head. It heals the very thoughts by healing the neurons, the nerve fibers, the brain cells. It makes the organs of application of thought lighter, more perfect.

Now, with each exhalation, allow the energy to return to the center of the room again. Your body is cleansed, energized, healed. Return the energy to the center ball again. Now, from this very healing center send out the words. Call those for whom you have concern. (Ed: Take time to speak their names aloud) Bring them in to enjoy this healing comfort.

Be at rest. Be at peace. Feel the energy expanding into every corner of the room. It not only has circulated through your bodies, it now enwraps your bodies and permeates your subtle bodies*. Even as the physical body is cleansed, energized, and healed, so are your mental, emotional, and spiritual bodies all cleansed, energized, and healed. So also is it with the loved ones who come to share with you.

As the energy spreads, and as it is shared, it does not diminish but grows in power. As the power increases, no longer can the room contain this energy. It must escape. It must spread out through the neighborhood, out into the town, and on and on and on, taking the cleansing, the energizing, the healing action into all corners of the world.

Bless* your efforts. Release your loved ones to return to themselves. Rejoice in the blessing received. And so it is done.
(Ref. S-23 - Sheariam)

---

*See Glossary

Daily Cleansing

(Purification by Breath)

To prepare, take in a breath slowly, and let it out very gently. Do that again two more times.

Now let your breathing return to its normal cadence.

Observe as you appropriate the air. As you draw in the sustenance of life, think about what happens to that air in your lungs. Think about the purification it brings into your body. Think about the waste that it carries away.

You should think about this as often as you go about your daily chores. The more awareness you give to the cleansing process, to the intake of air, and the exhalation of waste, the more your body responds to the energies of life.

You will have more vitality. You will find relaxation easier. Your sleep will be sounder, your dreams more meaningful.

The air comes into your lungs, and it is released. Life comes into your body, and it too is released.

Think about the cells of your body, each going about its own business doing its job, replenished by the life force you bring into your body with every breath that you take, cleansed and purified as it releases the waste material.

Let yourself go with this process for as long as you feel you need to do so. Let your entire body be cleansed, replenished, revivified.

When you are ready, refocus your attention to your immediate environment.

Bless your body for its service to you.

By placing your palms together in front of you, seal the cleansing process and the blessing.

Thank your body for its faithful service. And so it is done.
(Ref. S-19 Unidentified)

Body-Wise

(Energy Management)

Prepare yourselves for meditation by clearing your mind and relaxing your body...

As you relax your bodies, become aware of the sensitized points on your shoulders and your knees. Notice that there is a different perception there of the atmospheric conditions. While your body may feel warm all over, the points of the knees and the points of the shoulders may feel less warm. Or the reverse may be true. Tune in to that which it is for you...

Breathe deeply and rub the palms of your hands together gently. Then cross your arms placing your palms on the points of your shoulders. Feel there the difference in the energy of the body.

As you do this, you discover that the points of your knees also change, although you are touching your shoulders.

Now, release your shoulders. Bring your palms back together gently and, without crossing your arms, place your hands upon your knee caps. Notice the change in the energy systems of the body...

Bring your palms back together again, and simply relax. Breathe deeply... and rest your hands apart.

Using your dominant hand, while the other remains relaxed, make a clockwise circular motion with the flat of your palm over your midriff, about five circles. Then relax... Note the difference...

Now, with your non-dominant hand, leaving your dominant hand relaxed, make about five circles counter-clockwise over your midriff. Then relax... Note any changes.

These simple exercises should be practiced until, by their use, you can warm your body or cool it off. When you become emotional and your body energies become deranged, body temperature rises. Seat yourself quietly and perform these exercises until you have brought the body energies back into alignment.

When you then have the body energy systems properly aligned, you can proceed to your centering activities in preparation for meditation. This may help those of you who are in a spell of meditative inactivity.

Now, inasmuch as sound carries power, with your dominant hand, place your palm about three inches in front of your Adam's apple and say the AH OO MM* once. "AH OO MM." Now place the palm of that hand over your heart. Feel the power of connection.

When you are fearful, when you are frightened, apprehensive, or anxious, this will enable you to generate the energy of the renal glands.

---

*Refer to Glossary for "AUM"

Relax your hands. Now, with the dominant hand relaxed, place the palm of your non-dominant hand about two inches in front of the center of your eyebrows, and again say an AH OO MM. "AH OO MM." Now place the palm of that hand against the back of your head at the base of your skull...

When you need those words essential to convey a thought or an emotion, especially to someone with whom your relationship is disturbed, this exercise previous to such an encounter will awaken that wisdom energy in the pineal gland which you need to make communication at the psychic level and thus enjoy understanding.

For a moment relax and breathe deeply. Then placing your palms together before you, and, holding the palms together, again say the third AH OO MM. "AH OO MM." Now direct your palms to the center of the room.

You have not only aligned your physical energies, with the third AH OO MM you have brought into play the psychic and the spiritual energies which you now combine among yourselves to form a magnetic field in which rapport exists.

Relax your hands and allow that magnetic field to pull out of you that which no longer belongs -- release resentment, fear, lack of confidence -- anything which keeps you from being your total self. Allow that magnetic field to pull it <u>out of you</u> and <u>into itself</u>.

When you feel sufficiently cleansed, you may then, by speaking the names, invite those whom you think might also wish such cleansing to come into that magnetic field. Remember especially the absent members of your circle..... (Speak the names aloud.)

Affirm "Mercy and love are infinitely reproduced". (Repeat three times.)

With the creative power of your voices and the repetition of that affirmation three times, you have converted the negatives, drawn from you into the magnetic field, into positive attributes of mercy and love.

With the power of your thought, you may move that cleansing magnetic field into the atmosphere and around your planet to be partaken of by anyone who so desires. As that magnetic field moves across that portion of the Earth in which someone is asking for cleansing, in whatever way that petition may be presented, by magnetic attraction it is made available to them.

With the power of your thought visualize it as a huge balloon that moves through the atmosphere carried by the wind currents, so large that all of the sky is filled with it. As you do this, you bless not only your own planet, but the gravitational field of that planet, and thus the fields of the other planets.

(By participating in this meditation)....You have expanded your own capacity for further expression of mercy and love. As you move through the coming period of time be observant of yourselves. Notice when you have expressed that mercy and love which, being cleansed, enables you to express...

Thank those consciousnesses who have come to join you in this effort, and allow them to retreat back to themselves... Thank your own guides and ministers for sharing in this moment of cleansing and healing. Thank also the members of the Temple who have joined you in this effort, and who continue to join you in the times between your meetings... And then you may express your appreciation to your progenitor, to your God, for creating you in the image of a creative being which enables you to make a difference.
(Ref. S-54 Sheariam)

Focused Energy

(Chakra Healing)

Breathe deeply to relax your abdominal muscles. As you relax these muscles, the subtle bodies also relax... Move your attention through your body beginning at the crown of your head. Where you come upon tension, simply dismiss it. Simply tell it to leave...

Breathe deeply. Let your body release tension. Begin to visualize blue and green energy rays emitting from your abdomen. That is the creative center. As the creative center emits the blue and green radiation, you are sending forth healing energy.

These energies circulate throughout the room coming back into your body through your toes and through your fingertips. Let this circulation continue -- into your toes and fingertips, out through your creative center.

As this circulation continues, you become aware of a pulse beat. It is not a steady flow. It is a pulsating thing. Allow yourselves to flow with the rhythm of the cycle.

Now gently but easily stop the flow from your abdomen. Allow the energies to come back into your body and direct them to any portion of your body which is disabled in any way. As the energy is directed by your conscious will, you will feel that part of your anatomy becoming warmed.

If you wish to heal another kind of disablement, a relationship, feel the warmth in the heart chakra. If it is a matter of courage, or decision, a matter of personal power, feel the warmth aggregating in the throat chakra. If it is for illumination, understanding, feel the energy aggregating in the third eye area, the center of the forehead....

Now accept your personal healing. Become at ease with that energy. Bless that part of your body, or that relationship, or that situation...

Allow the energy to move down to the pit of your stomach. This is the point of your anatomy which deals with peace. Let the energy rest in that chakra. Feel the energy systems of your body becoming balanced....

Now as you become peaceful inside, you can release that energy again as peace. Let it move from the center of your peace to join with all the others in your midst, forming a great ball of blue and green energy that is organized in the principle of peace.

Before you release this energy to the world, take a moment now to include your loved ones. Invite them to join you by speaking their names aloud...

Now in peace and goodwill bless those who have come.. With your thoughts of peace and goodwill go forth your thoughts of blessings. The great ball of peace energy is lifted from your midst and sent aloft. The energy is dispersed through out the heavens and around the world...

Take joy in the peace you feel within yourself. Allow your loved ones to take joy in theirs and return to themselves.

Now bring your own consciousnesses back to daily awareness. Go in peace.
(Ref. R-5 Sheariam)

Relaxation And Self-Affirmation

Settle yourselves into a comfortable physical position. You should have your body sufficiently comfortable that you do not have to pay it any attention. Of the first consideration is relaxation, so before beginning the meditation, you may wish to do the following relaxation exercise:

Take three deep breaths and exhale through your mouth. Then return to normal breathing... For a moment observe your breath without your conscious will. Your breath comes into your body, replenishes your blood stream, carries away the toxin and leaves your body. Divine Will has provided a means by which your body is replenished and cleansed. Just objectively observe...

Now let your body muscles relax. Clench your fingers and toes. Then let them totally relax. Feel the tension drain away... Now tighten the muscle in your shoulders and your thighs. Then relax and feel the tension drain away...

Now contract the muscles in your abdomen. Then totally relax and feel the tension drain away... Now swallow hard. Clench your teeth and squint your eyes together. Then relax and feel the tension drain away...

Now continue to be aware of the tension draining away, moving away from you. In your thoughts move away from your body. Observe it with detachment. Recognize that part of you that is not your body... Think about how you feel right at this moment... Happy, sad, whatever... and move away from that feeling. View that feeling with detachment. You see you are not what you feel...

Now examine what you are thinking, and separate from your thought. You see you are not your thought. So there is that portion of you that is detachedly observing -- observing the body that you wear, the feelings that you carry, the thoughts that you think, and yet is none of these things.

As you breathe normally, each exhalation of breath breathes out "I AM". Just be still and let that be the consciousness -- "I AM". If you are having any problem, remind yourself, "I am not my body; I am not my feelings; I am not my thoughts." "I AM"....

Now to move back towards the current reality, affirm aloud:

"I AM A CHILD OF GOD"... "I AM A CHILD OF GOD"... "I AM A CHILD OF GOD"... "I have a body for which I am responsible. I bless the body I have... I have feelings for which I am responsible, I bless my feelings... I have thoughts for which I am responsible, I bless my thoughts... I am a child of God and I am blessed."

When you are comfortable return your consciousness to this room bringing with you that feeling of blessedness. Move your body. Press your hands together to lock the energy within.... and SMILE.
(Ref. T-7 Silverthistle)

Elastic Arms

(Mending the Aura)

Settle your bodies comfortably and begin to focus your attention on your breath. Breathe in deeply through your nose and exhale through your mouth at least three times. This assists in the relaxation of the chest and abdominal muscles....

Then slowly, in rhythm with your breathing, allow your attention to focus into the various parts of your body. Just become aware of the body. Feel the feet and how they touch the floor. Feel the legs as they rise from the feet. Just become aware.

Feel the flex of the knees and the feeling of the thighs against your chair. As you focus your attention into the body, wherever you become aware of tension, just dismiss it. If you have a problem, flex the muscle and then relax it...

Continue up the torso. Become aware of the arms, elbows, hands, throat, neck, face, head, until you are completely aware of all your parts, but relaxed.

Then allow your consciousness to expand beyond the limit of the skin of your body. Become aware of the vital energy that surrounds your body. These are your energies. They, too, are a part of you. Your breathing remains gentle and rhythmical.

As your consciousness expands you begin to feel lighter, and a sense of joy begins to permeate you, for as you expand your consciousness into the subtler realms of your body, you become aware of greater things to be experienced, to be enjoyed....

Your imagination is the greatest faculty you have for exploring the finer realms of your being. With your imagination, extend your arms beyond the limits of your hands. Allow yourself to let your arms be elastic so that they become lengthened sufficiently long that you can reach around yourself to any part of your body.

As you have examined your body and become aware of it, you discovered areas in the body which were less comfortable than other areas. If you have a particularly tender spot or some such, with your extended hand soothe that spot. Massage with your extendedness any part of your body that may require attention.

As you do this you are mending your auric field. This is something you can do any time. When you have finished soothing your aura, mending it, mentally invite your personal healers to serve you. It is not necessary that you name them, or that you can even be aware of an individual personality. Just open yourself to that loving service, that loving ministry, and your healers will come to you, touching you where you need their healing. Just relax and let it happen...

Even as you are receiving treatment, you may find certain symbols appearing in your mind. These will be clues to the identity of your healers....

Now that you have received some reconstitution, some alignment has been made for you, share these energies. Allow yourselves to glow. Allow the inner light to emanate from you and join the light of all others in the circle. So you become united as you flow together in your subtle energy.

You may think you are only exercising imagination. Do not let that 'only' dissuade you. Even as you pour forth your light, the healers and other members of the bands gathered with you emit their light, and the room becomes filled with their light -- the healing energy, the teaching energy, the caring energy.

As your light mingles, there is even a heat generated. Those of you who have the sight will be able to see this energy in shades of blues and greens and irridescence.

Now one of the very beneficial things you can do is share this light energy with those who are of concern to you. As you speak a person's name, there is a connection made even though he or she is absent. Speak the name of your loved ones and let that call their consciousness forth if they be willing. As that consciousness is called into your midst, they, too, benefit from the healing energy. It is best if you speak the first name aloud.

It is important that you share these experiences with your loved ones even though they are not present in their physical bodies. Through the sharing of these experiences, you develop, you grow, you attain to that altitude at which you not only receive a blessing and a healing of your own, you pass it on. As the healing goes forth through these other consciousnesses, the healing is spread forth from them, too.

So you share. You make a contribution to the elevation of consciousness throughout your realm. And your entire earth planet is benefited.

Now in your visualization allow the brighted, lighted energy to come together into the form of a fountain in your midst. Let the fountain grow higher and higher, until you are re-bathed in the waters of your own healing light. The stronger you can visualize the fountain the more forceful the energy becomes.

Then let the fountain grow higher and higher until it spreads out and indeed waters your entire globe, taking the healing to the farthest reaches of your planet. So you bless your planetary home as well as your temporal home.

When you are ready, bring your awareness back to this room. Move your body gently. Clasp your hands together in front of your breast to lock within your body the healing you have received. So you are blessed. Go in peace.
(Ref. R-2 Sheariam)

Foot Massage

by Temple Healers
(Understanding and Flexibility)

Breathe deeply to generate oxygen content within your body. Load up your systems with oxygen. Put your consciousness into your feet beginning with your toes. Allow the oxygen-laden blood to circulate warmly through the feet. It is imperative that you keep the oxygen content in your bodies at as high a level as is possible.

*(Await quietly the arrival of the Temple Healers)

Direct the consciousness into the feet. Let the healers' hands begin to stroke your insteps and pull at your toes. Feel your toes individually manipulated and massaged. As the bottom of the big toes are massaged in a circular motion, you find the tension withdrawing from your neck and shoulders, and your scalp muscles are relaxing.

The soothing strokes now being administered to your heels and down the sides of your feet relaxes you further.

Then, as they massage on the soles of the feet, on the balls of the feet, the energy there begins to whirl, begins to tingle, and yet you are deliciously relaxed... Now, each one of your

---

*We have been assured that when we sit in these meditations and ask for the healers, they will join us even when the Temple Teacher is not present. For further instruction on this see Appendix.

healers will manipulate your feet in accordance with the need of the individual. Adjustments of the tiny bones are made; the ligaments are gently stretched and exercised. Where there are growths upon the bony tissue, they are gently massaged away. Where there is misalignment, the feet are placed in alignment again. Just relax and enjoy the ministry.

As the minute adjustments are made, you may feel a little discomfort. Know that no injuries will ever be inflicted upon you. The discomfort is due to the fact that the feet have been in misalignment for so long. True alignment brings some degree of discomfort at first.

Now if you will each simply think to yourself the number 7, the energy of that number assists in making the adjustments hold.

Now the healing moves on, up into the ankle. Allow your consciousness to rise to the ankle area. As the feet have to do with your understanding, your ankles have to do with your mobility. If the ankles are not supple and flexible, one tends to have inflexible thought processes. So even as the massaging fingers move to adjust in the ankles, so the spiritual forces move to make adjustments in erroneous opinions and thought patterns.

Wherever there is the weakness in the ankle, the healers put in energy to strengthen ligaments, muscles, bone tissue. Again, in the moving joints there are emollients applied to make the movements more fluid; to dissolve any unnecessary bony spurs.

Even as the ministry continues, you find that your facial muscles relax. There is less tension around the eyes, around the nose and mouth... Inflexible thought patterns cause tension in the facial muscles.

Now the healers will conclude their ministry with upward strokes to the knees, increasing circulation, increasing eliminations of toxic wastes.

Now, each in his own way, thank your ministering healer. Allow your consciousness to return to encompass all your parts simultaneously. Bless your body. Bless your Self. Go in peace. And so it is.
(Ref. S-26 - Sheariam)

Triune Balancing

(The Body's Inner Knowledge)

Settle your bodies. Relax your minds... Relax your eyelids -- squeeze your eyes together, and then relax them... Compress your lips, and then relax them... Swallow several times... Then breathe deeply several times...

Now, as you allow your breathing to return to its normal cadence, just observe. Be aware of that inner knowledge of the body itself to perform its functions... The lungs billow, and the digestive tract moves in its own rhythm. When you need them, your muscles tense for use. When you tell them to relax, they know they are not needed. But those muscles that are needed to keep you upright in your chair remain on the job. The muscles necessary to keep your head on your shoulders remain on the job.

Think about your body as the marvelous instrument that it is...

Now think of the emotions which act upon that body, to swing it this way, or move it that way... Step aside and look at the thoughts which influence the emotions...

Then look at it from the other perspective -- the emotions which influence the thoughts...

Then go back to observing the rhythm of your natural breathing. With each intake of breath, allow the mind, the thoughts of the mind, to line up with the emotion, to become balanced with the body so that you have a triune balancing of mind, body and emotion...

It is the balance of these three which enable the physical organism to be whole and effective as a vehicle for your finer self to live in this physical world.

Allow your consciousness to rise into that finer self of intuitive knowing, as though you stand above and behind your body...

From this higher elevation of the intuitive self, extend your palms toward your body. Bless your body. Bless your emotions. Bless your thoughts...

From this intuitive elevation, you can expand without limitation of the physical. Allow yourselves to expand toward each other until you stand in a joined ring clasping hands...

Behind you, but near, the healers from the Temple stand with joined hands. Energy moves around and around the circle, passing through your hands to each other. Energy passes through the healers' hands to each other...

These two circles of energy act as a dynamo for healing... This energy rises from the two circles of people, coming together above them into an umbrella configuration of energy, healing energy. As you look up at it, you can see the iridescent greens and blues of the healing colors...

While you maintain that umbrella with the assistance of the healers, invite those loved ones who may wish to join you to come in under that healing umbrella.... (Speak the names aloud.)

Now, even as those who have accepted your invitation continue to join you, see that umbrella of healing energy move downward, covering you, permeating you, enlivening you, warming you. Accept the healing. Feel that thrill of joy within, that inner knowledge of contact made.

In the act of being healed, you are loving the healer. You are loving yourself. You are being beloved...

As the healing energies mix and flow with you, you mix and flow with them. You become one with all. The consciousness of the group becomes one consciousness, which rises... and it rises...

The love you feel for each other rises with that consciousness and becomes a blessing to all who would partake of it. Allow your consciousness not only to rise, but to expand, carrying this healing energy and your love to all areas of your planet.

As healing takes place, peace must ensue. Abundance is provided. Intelligent decisions are made. Feel the contact that you make with people you will probably never even see...

Now release any claim that you have on that energy, knowing that you may have all that you need without depleting the source. Begin to draw your consciousness back to the limitations of your body. Bring your awareness back into the physical body, into your present condition, into the awareness of this room. With blessing and love, release your loved ones to return to themselves. Thank the healers of the Temple who have come to join you in this moment of blessing the Earth...

As you bring yourselves back to daily consciousness, go in peace.
(Ref. S-51 Sheariam)

Blessings Return Ten-Fold

(Cleansing by Transmutation)

As soon as you have settled in a suitable position to attain the most relaxation, take three deep breaths into your nose. Exhale through your mouth. Then allow breathing to return to normal cycle. On each inhalation say to yourself "rest". On each exhalation say to yourself "relax". "Rest" and "relax". "Rest" and "relax".

As you say these words to yourself, you are beginning a re-programming. Hereafter, whenever you breathe in, there will be a signal to rest, and when you breathe out, there will be a signal to relax. (You are) not programming to sleep, but merely to become calm. As you say these words your breathing becomes slower, and your body responds with relaxation.

Now in your mind detach yourself from your physical body. Simply observe the breathing cycle. Watch it.... Watch it... Watch it with objectivity. Feel it for the miracle that it is.

This vital life-giving process continues for you day and night without thought. Charged atoms of oxygen are drawn into your lungs to enter your bloodstream, and like iron magnets draw to themselves the toxic waste, carrying it out away from the body by the lungs through exhalation.

From this detached point of view, observe these atoms of oxygen with their payload of waste, moving out of the physical body into the desire body, into the emotional body, and passing away from the physical. As they pass through each of these energy fields, certain of the toxins and wastes are transmuted so that by the time the atoms of oxygen are entirely out of your physical field, it is again clean and ready to be recharged, in order to again perform the cycle.

Now re-establish yourselves by repeating "rest, relax" in rhythm. Know that you can bring healing to your physical organism by constantly thinking about the charged atoms of oxygen as they are drawn into your lungs. Visualize the cluster of oxygen atoms clean, highly charged with life-giving vitality. They are drawn into your lungs and blood stream, there to perform their cleansing work. Visualize them as though they were little bubbles in your bloodstream drawing to themselves all those toxins and waste which are inimical to the health of your organism. The more powerfully you can visualize this, the more quickly you can be healed.

When you get this cluster of atoms nicely loaded up with the undesirables, visualize them being taken back to the lungs, expelled into the air, and being expelled from your body. Do not stop the visualization there. Carry on with these loaded atoms and see them transmuted. See the toxins and the waste transmuted by chemical action until the atoms of oxygen are again clean and available for service. In that way you not only bring blessing into your body but you actually also bless its eliminations, turning them into positive energy.

Now in the same way, you can visualize a thought as an anatomical particle which comes into your mind, gathers negative thoughts, negative feelings, pain, harrassment. You can send that thought back away from yourself, transmuting the negative that thought has picked up. All these negative energies can be transmuted by the subtle body into positive love energy. Then that thought is cleaned and ready to be recharged into a vital, positive, loving, blessing thing.

Develop in your consciousness the act of blessing everything that your eyes fall upon, every thought that comes to your mind, every word that goes forth from your mouth. "Bless you, bless you". Under the universal law of ten-fold return, each blessing that you emit, returns to you in ten-fold blessings.

Now in this quiet relaxed state, think of someone who represents painful experience to you. Bless them. Even as you think that thought, "bless you", the very negative within yourself that you found difficult to get rid of, disappears. With repetition of this exercise, you will relieve yourself of much negativity. These very people become a blessing in your life.

Now bless yourself, and bless the effort you have just made. Bring yourselves back to the consciousness of this room. Prepare yourselves to generate the physical energy to send out the healing rays.

When you are ready, rub your palms together and generate the blue and green lights. Let them come together as a clover leaf. Then rest your efforts, and invite your loved ones by speaking their names aloud...

See the clover leaf become a three-linked chain of fidelity, truth and love. Visualize an arch over your invited ones -- a blessed arch of protection. See the blue and green lights like petals at your feet and the three-link arch like a rainbow overhead.

The petals grow larger, dissolving into swirling lights, moving out, including each one in the circle. Again rub your palms together, and then turn them to the sky. Open your palms, energize even further the blue and green so that it circles your globe bringing peace, healing, faith, love and truth to each and everyone who will accept the energy.

Bless these energies as you send them out, intensifying the power. As you send out your blessings, release your loved ones. Bring your palms together locking within yourself the healing energies received. Now bless yourself. Go in peace.
(Ref. T-11 Silverthistle)

Piece Of Gauze On God's Clothesline

(Whirlwind of Universal Cleansing)

Prepare your bodies. Prepare also your hearts and minds. Relax. Be porous. Be receptive and be willing to allow the energies to pass through you -- do not stop them within you. (When you stop them within you) you become an object in your own way and in the way of others.

So visualize yourselves as a piece of gauze through which light filters. Open yourselves to the passage of the healing energies... Permit yourselves the fancy that you are a piece of gauze on the clothesline of God...

The light shines from the heavens and the breeze passes through you. You flutter with the breeze. The sunlight cleanses you, and the breeze carries away anything that needs to be eliminated from your texture...

You are willing to flow with that, but you are always anchored. You are anchored in that connection of God's clothesline... Thus it is that the winds of life blow through your experience. You can allow them to pass through as you remain true to your own nature. Your impurities become bleached by the sunlight of God. You remain ready to be gathered... when the time is right...

Now, as a fragment of the Divine texture, allow the healing energies of the Universe to pass through your minds, through your hearts, and through your bodies... As it passes through you, it goes on to pass through others.

It is that eternal radiant energy emitted by The Absolute in the Creation... Feel yourself cleansed, purified.... As this energy continues to circulate, it builds a power which then becomes available. Visualize this passage of energy gaining in power even as a whirlwind swirls its way along, becoming larger, larger and more powerful...

This whirlwind is not one which gathers up debris, but remains the pure energy of The Absolute, the radiant energy, the Truth... As you witness this growing power, you become aware of it in your environment and in yourself...

Since it passes through you, it is available to all. You may invite those about whom you have concern to join you in the experiencing of the whirlwind of loving healing. (Speak the names aloud.)

The whirlwind gathers up from you that which you do not need, eliminates congestion, obstruction, all that is unnecessary and whirls it away into the universal cleansing where it can be recycled. So release, release... Let go, and be cleansed. Be renewed...

As you are renewed, see the whirlwind carry away the negative energies of the Earth planet to bring change into the thoughts of mankind, to deliver peace and love, to purify this planet and this condition...

Release your loved ones to return to themselves. Know that at any time you can stand within the holy whirlwind and release to its cleansing energies, and be healed... Now go in peace. (Ref. S-49 Sheariam)

Charging The Chakras

(You are a Creator)

Relax your bodies and relax your minds. Focus your awareness on that top-most point of your head, the entry point of universal energy*... Visualize the plates in your skull as movable... Because all things are possible in your imagination, pretend that they are in some way controllable, that you can slide them apart from each other, thus opening your head to the influx of universal energy...

If this immediate inpouring becomes uncomfortable because you have not been used to taking in all of your supply, you may close up those bones in your head whenever you are ready. If you can sustain the influx of energy without discomfort, leave them open for a while.

Be aware of the energy as it enters... as it passes through that midpoint in your brain, which is called the brow chakra... This is the point at which the energy is polarized into acid or alkaline. If you have been suffering with indigestion, then think of the alkaline polarity. If your body has been irritatingly dry, think of the acid polarity.

Then let the energy pass down through the center of your throat, through the throat chakra, where the thyroid gland again divides the energy. One portion of it becomes anabolic and the other portion becomes catabolic. Some of it moves down your arm to linger in your fingertips. But the greater portion of it becomes resynthesized in the heart chakra, in the center

*See Glossary

of the chest. Your body may become very warm as this occurs.

Then the energy flows down into the solar plexus and becomes physical energy; that is, into the form in which the physical organism utilizes it. Then the energy passes down into the center of the abdomen and your reserve battery fills up. The rest is delivered to the root chakra at the base of your spine. Here it has become red energy, life promoting energy, creative energy.

It is apportioned into the legs, goes down the legs to linger in the toes. Now as the circulatory systems work, the lymphatic system, the blood system and the energy system, the used energy is drawn up again from the toes to the root chakra.

This is the moment when your creative mind functions. Make the decision as to how that energy is to be transmuted by you before its return to the universal. With your creative thought draw that energy up into the abdomen. There your creative thought begins to take physical form. Then the energy is drawn up into the solar plexus, and out of the solar plexus the thought form which you have created is emitted to find its substance and reality in your world. It now has force of its own.

The life promoting energy is then drawn up into your chest where it is transmuted back into what you call spiritual energy. You feel love for your created thing, thus giving it some of your electrical charge.

Then you draw that energy back up into the throat chakra. Here you give the thought of power to your created thing, and it is empowered to become alive.

Then you draw that transmuted energy out of the throat chakra into the center of the head, into the brow chakra, and here you bless your created thing. You imbue it with goodwill, with good intention, and direct it to serve its purpose fully.

Now having made a contribution, you pass that transmuted energy back up out of the crown chakra, at this time drawing up the energy that has lingered in the fingers. Even as this energy is passing out of you, other energy comes in. It is a constant flow.

Allow the head bones to return to their normal position and know that at any time you wish to create something, you can do this same meditation.

Now with your mind set, deliver some of your charge to the center of your circle. Allow that force of your electrical charge to mingle with the force of the electrical charge all around you. As your auras interpenetrate, you bring balance to each other. You bring balance to all the energies in the room. Add to the power of those mingled forces your thoughts of love.

With the power chakra, with your voice, announce the names of your loved ones in invitation that they may bring their auras here to share in this balancing action.

Feel within yourselves the balance, the readjustments. As you feel that balance within yourselves, you are aware of peace, of equipoise. As you are peaceful within yourself, your creation also experiences peace.

With those who have come to join you, expand your auric field. Allow the thought of peace to be transmitted. The farther you can expand your auric field, the farther you can trasmit your thought of peace. With the combined charges of these fields the message becomes very powerful.

Now release your loved ones to return to themselves. Draw in your expanded auras so that you will be protected when you leave the group, but retain within your heart that sense of peace, that sense of balance. Be ready for the consequences of the manifestation of your creation.

Bring your awareness back to this room. Be blessed. Go in peace.
(Ref. R-8 Sheariam)

Circulation of Divine Energy

(Finding Answers)

Make yourselves comfortable... Align your head atop your spine. Breathe a number of times deeply to help your body relax... Place your palms a short distance apart and begin to build the ball of energy as you have been taught.

Look into that space between your palms and see the blues and the greens of the energy ball as it forms... As the ball grows larger, allow your hands to grow apart until you have a ball of energy about 8-10 inches in diameter... Then flatten your hands and blow that ball of energy into the center of your midst.

As your lifegiving breath gives impetus to the ball of energy, all of them come together in the center of the room, and like solid objects falling into a pond of water, they make a fountain of splash rise. Visualize that fountain -- the blue and the green waters rising and splashing.

As the waters rise, feel that triangular space* in the crown of your skull begin to open... Allow this energy fountain to splash some of itself into that opening.

As the energy passes through that opening, it moves down through your head into the top of your spine; then on down through your spine. From there it transmits itself throughout your body. Simply let it flow in your head, through

---

*See Glossary for "fontanel"

your body, and out your toes and fingers to mingle again with the pool of energy in the center. Be bathed in that healing energy. Be warmed by the divine love it represents. Be filled with the divine fire it represents...

Because this is the divine energy in circulation, you may ask whatever question you hold in your consciousness. As this divine energy passes through your head, it will stimulate within you the answer...

As healing comes, so do the answers. As the answers come, so does the healing. If there is irritation, nerve ends are soothed, washed cleaned, anointed.. Where there is fear, confidence is established... Where there is pain, calm is granted... Your whole body feels washed and peaceful.

Into this healing, blessed realm of experience, invite those about whom you have concern... (Take time to speak the names aloud). As your loved ones come to join you in consciousness, visualize the fountain rising higher and higher so that the healing waters of peace extend beyond this room to your city, to your nation, throughout your planet. As you visualize this, you also affect the Realm of Spirit.

Allow the waters to rise even to the limitless space... Now allow the crown of your head to close, holding within yourselves the blessedness of the healing. Let the fountain dissipate into the ethers, carrying its blessing wherever consciousness calls. Release your loved ones to return to themselves. Be blessed. Go in peace. (Ref. S-43 Sheariam)

Ball Of Energy

(Mercy, Love and Compassion)

Begin by rubbing hands together in the following manner. Put your palms together. Then with a circular motion, rub them slowly, almost sensually, until you begin to feel some energy building. Then, little by little pull your palms apart, still maintaining the rotation motion. When your palms reach about 4" apart, you will feel as though you are rolling a ball of energy between your palms. Do not do this briskly, for this is a very controlled exercise.

Breathe deeply as you do this. Breathe in through your nose and out through your mouth. Relax your abdominal muscles. Release all the tension in your body. Particularly release any tension you feel in your forehead, in your eyelids, or any other facial muscles. Keep the hands moving and begin to separate the palms. Now concentrate on what you are doing, but do not be tense about it.

Very gradually, widen the area between your palms. Feel that ball of energy. Now that ball of energy is as fragile as a bubble -- treat it as such. But it is very powerful -- do not be misled. Within each of you, there is the quality, the capacity for mercy, for love, for compassion.

As you breathe in through the nose, breathe out through your mouth so that your breath falls on the ball of energy. Allow feelings of mercy and love, of compassion to be breathed out -- out of your own being and into that ball of energy. It becomes more powerful; it becomes more brilliant; it begins to pulsate. Indeed it begins to have a life of its own.

In a moment you are going to waft your ball of energy into the center of this circle. Like bubbles, they will cling together to form one large bubble. Now very gently breathe in through your nose, and as though blowing a feather off the palms of your hands, blow your bubble of energy into the center of the room. There see the energies combine. One huge, beautiful, iridescent bubble of energy grows larger and larger. There is unlimited capacity here. There is no crowding -- everyone is welcome.

(Take time here to invite your loved ones to join you.)

As the bubble shimmers and moves, it radiates energy, and that energy comes back to you. You feel this iridescent energy as it brushes the palm of your hand, and the combined mercy and love and compassion of all present in this room now becomes part of you. Each of you take what you need.

In your physical existence, there are times when you may get a certain substance upon your hands, and it's absorbed through the skin. Soon you can taste it in the saliva of your mouth. Allow this wonderful, healing, iridescent energy to permeate the skin of your palms. Let it be carried through all of your body until you have a sweetness in your mouth. Your own reservoir for mercy, for love, and for compassion is filled and enriched, as you partake of the blended energies.

As you brought your loved ones in mentally, mentally return them to themselves. Clear the ball of energy of all human occupancy. With the power of your thought, waft it away from this room. Send it around the globe. Let it travel through the atmosphere. It will give off the energy of your combined mercy, love, and compassion. Thus your own beingness extends its blessedness to each and every other occupant of this planet. People are healed. Situations are healed. Conditions are healed, and yet you are replenished. Bless the ball of energy. Let it continue, let it go.

When you are ready, take a very deep breath. As you exhale bring your palms together locking the blessing within your own form. And so it is done. You are very blessed.
(Ref. S-20 Sheariam)

Cycling Energy Through You

(Working With Divine Germ Cell Components*)

Settle your bodies, and put aside the thoughts of recent moments. Clear your minds and your hearts. Relax your bodies. Breathe deeply...

Envision a circle of light coming in through your crown, and moving out through your toes..

As it moves out through your toes, it moves out into the Divine Spirit, gaining thrust to enter the crown again... cleansing the body as it moves through the body and out through the toes... to again be purified and strengthened. Give yourselves a moment to experience this cycling of energy, through you, through the Divine, and back to you...

Now, as you experience this, become aware of the two forces -- the inward and the outward... Feel the pulsation.

Then become aware of the percieving mind and the creating mind. And all the time, the circulation continues...

Then become aware of the increasing vibrancy within you that is the life itself...

Then think about your quality of humanness. And the circulation continues...

---

*See Glossary for "Divine Germ Cell Components"

Give some thought to the quantification which you have put to this quality of humanness... In this moment, set some new goals of increased achievement as a human being...

Now turn your attention to volition, your capacity to act upon those goals...

Contemplate consciousness as being that focus of energies which you are. And the circulation continues...

Now find that center within yourself that is the center of the love consciousness. Let the feeling of love flow from that center out through your hands to the center of the circle. Let it there form a great pool of love...

Then take for yourselves what love you have need of, knowing that there is all the love necessary to serve all the needs of everyone here as well as the needs of those for whom you have concern. Call forth the names of those whom you would invite to participate in this pool of love...

The healers from the Temple now bless your efforts, adding their own energy of love and blessing the consciousnesses of those who come at your invitation....

Release those whom you have called to return to themselves. With the power of your thought, disperse the love energy throughout the planet. Make it available to all who would partake, for where there is love, there cannot be war. Where there is love there must be abundance...

Know now that you are baptized in the waters of love. And with your thought, you cleanse the world. Be blessed. Go in peace.
(Ref. S-48 Sheariam)

Sea Of Energy

(Dissolving Cares)

Be at ease. Relax yourselves... Release the thoughts of your mind. Release the tensions of your body. Relax the emotions. Let go your cares. For the next few moments, be highly focused.

Visualize a rod of electric blue light rising in the center of your circle... Then let that rod become like a column of electric blue water which flows outward from the column forming a sea of electric blue energy...

Float in that sea of energy, carefree, relaxed, buoyant. Feel the buoyancy of the energy, like that of water.

Unlike the water of your planet, this field of energy has no capacity to act upon your body in a negative fashion -- you are safe, secure, and held lightly...

Allow this sea of energy to wash through you as well as around you. Allow it to cleanse you... Allow yourselves to feel dreamily peaceful...

Now, in this great solvent, allow any concern to float away, to be dissolved. Be objective. Detach yourself from the concern. Know that this universal solvent can dissolve it out of your life while you remain dreamily peaceful, intact, relaxed and buoyant...

This is one way in which you can enact the Law of Relinquishment*...

---

*See Glossary

If you need to, use your powers of visualization to place the object in this sea, and let it float away from you, knowing it will be dissolved never to return...

Take delight in your freedom from that concern. Take delight in your healing, in your well being, in your sense of comfort and security....

Now, in your knowledge that something delightful shared increases the delight, invite those of your loved ones whom you would wish to share in this healing sea. (Speak the names aloud.)

Be at peace. Know that those who respond to your invitation will find the same cleansing that you find. Invite them also to cast their cares away from them, into the solvent that they shall be detached and cleansed and healed ...

Unlike the waters of your planet, this sea of healing energy takes what is put into it, dissolving it back to primary substance. Thus, there is no pollution. That primary substance is once again available for use. It is no longer imprisoned in the illnesses, the fears, the concerns of this Earthly life...

Like the sunshine which evaporates the moisture from your Earth, know that God-love draws this sea of energy into the atmosphere. Like the sun burning away the clouds, you find yourselves free in the Divine Light, cleansed and whole. When this is accomplished fully in consciousness, you will experience what is called the Resurrection...

In your cleansed bodies, in your cleansed minds, your cleansed emotions, in your wholeness and purity, allow that sea of solvent to move through the atmosphere as clouds bearing moisture to be delivered to those areas of the planet where demand is made for it. Visualize it as a great body of cloud moving through your atmosphere and around your planet...

As you move this with your thought, those consciousnesses who are ready for such cleansing will draw upon that power, even as you have in times past, though unknowingly...

Bless your loved ones and allow them to return to themselves, even as you bring your consciousness back to this daily realm. Carry with you that sense of security and cleanliness, the relinquishment of cares to the eternal solvent..

Thank the Temple healers and all the helpers that gather with you who give you support, protection and love. Go in peace, with love. (Ref. S-57 Sheariam)

I Am

(Love of Life)

Settle your bodies. Find that center of focus within yourself wherein you know "I AM".... In that center you know you are not your body. You are not your mind nor the thoughts of your mind. You are not your emotions nor the feelings of your emotions....

In the next few moments, as you breathe in, think "I AM". Then release your breath... On the next breath, "I AM"... As you build that consciousness within, the power of that awareness draws to you the other subtle bodies. As your extension becomes more vibrant, you begin to glow from that center of I AM.... The light of your beingness joins the lights of all the others.

From that center of the I AM, imagine going forth from your solar plexus a ray of vibrant light that is your love of life itself.... As you love life, you know from that I AM center, that that love and that light is not dependent upon the physical manifestation, nor on the subtle manifestations, but indeed only dependent upon the Divine I AM....

So in that calm center of I AM from which issues the love of life, there can be no presence of fear. There can only be the certainty of Divinity. Regardless of what may occur in the condition of the physical, or any of the other conditions appertaining to the physical, the I AM is eternal...

With this certainty of eternal life carried on the wave of your love and the vibrancy of your feeling, invite your loved ones to join you in this realization... (Take time to speak the names aloud.)

Where love abides there is perfect health. Let that feeling of loving life move throughout your physical organism. All cells respond to love, and the order of love is established within your being. In that orderliness you can only be healthy, prosperous, and happy...

With a fullness of love, release your loved ones to themselves. Bless them as they return to themselves. Spend a moment now blessing your own body from head to toe...

Bless your life from its innermost parts to its outermost extensions...

With these blessings there arises within your heart a sense of gratitude for the Divine plan that enables you TO BE.

Give a thought of thanks to your higher self*, to your guides and ministers and teachers who serve you silently and faithfully at all times...

Go in peace. As you go in peace, those with whom you meet go in peace. And as the message spreads, so you do what you are supposed to be doing, and all of mankind benefits, as does the planet on which you live.
(Ref. S-47 Sheariam)

---

*See Glossary

Celestial Music

(Influence of Sound)

Relax your physical bodies. Be at peace. Know that from the Realm of Spirit come melodies of celestial music, which, even though they may not be heard by the human ear, nevertheless occasion sensations in your life. To whatever degree you find possible, open the center within your breast, that seat of emotion, and like an empty cup receive this heavenly melody.

Let it fill you. Feel the vibration. Allow it to rock you as a mother rocks a child... Feel how it soothes, quiets and calms you... Visualize this melodic infilling as light...

The chalice fills and reverberates. It gives off beams of light. The calmer, more soothed you are, the greater is the intensity of the light you emit...

Visualize this glow moving through your subtle bodies and enhancing your auric light.

Throughout your subtle bodies you are utterly luminescent with this celestial melody...

As the light moves outward from each, it joins the light emanating from the other, and the room becomes filled with light and melody...

The light grows brighter. Your body feels lighter, more porous. You feel very expansive...

Into this healing vibration of light and melody, invite those about whom you have concern. (Speak their names aloud.) Instruct them to absorb, to drink in and to enjoy this celestial harmony....

Accept the healing within your own body as you respond to that beat. You become aware as humans that you have a common melody. You share in a community of harmonic beingness. Rejoice in this...

Allow this melodical light to disperse itself throughout the atmosphere of your planet, bringing life and harmony to all who will accept... Know that even as you go your separate ways, that connection still sustains. You are still melody-filled. You all sing the same song.

Bid your loved ones farewell. Conclude your meeting with your own song of peace.
(Ref. S-59 Sheariam)

Peace

(I-Am-ness)

Relax your bodies... Breathe deeply so that your brain cells have lots of oxygen. Become aware of yourself as a selfhood.

Know yourselves as a physical body, as an emotional being, so that your heart beats with love, with yearning, with on-goingness...

Know yourselves as a mental being, that you can think. Then also witness that you think. And in that mental body you can think of things to do and to be and to create... In that mental body you can make plans, and you can think about yourself making plans...

Still beyond that, there is that sense of I-AM-ness that is not merely the thinker, that is not the feeler, that is not the gross matter that you know of as physical body...

Then there is the I AM that knows that it is I AM....

In reaching toward that sensitivity of eternal I-AM-ness, you activate that celestial substance which is that eternal rod of life. That star seed which has become you has viability, has significance, has great importance in the scheme of things...

So you touch that part of your soul that makes you know responsibility for what you create.

But the I AM is beyond that ...

That brings you to touch that part of your soul that enables you to know that there will never be a time when you will cease to be....

Out of the depth of deep, deep darkness comes that shaft of light of understanding that calms the fearful heart, that heals the wounded spirit, that brings out the courage that cannot be lost because of the eternal urge to be.

And in that, you know your God Self.

In that God Self, you know your indestructibility. From that center of knowing, you emit rays of warmth and light, of love and healing. You make yourselves whole.

As you continue to emit the celestial light, which is the very substance of you, invite those who will to join you in this moment of wholeness and love and peace.... (Speak the names aloud.)

Know that the glorious light of your beingness joins with theirs. You become one in mind. You become one in heart.

You become one in purpose. Together you choose the purpose -- peace. In that purpose you know lies the wholeness for this Earthly home.

With the power of the mind you send this vibration of peace, this condition of peace around your planet. It circles and enshrouds the planet. The planet becomes encased in the light of your thoughts and the warmth of your hearts. Your planet becomes healed and whole.

So you have not only served yourselves, you have served your loved ones; you have served us; you have served this physical realm which we all inherit...

"Beloved, go in peace".
(Ref. S-66 Judith)

Three Sided Golden Pyramid

(To Open Psychic Centers)

Settle your bodies comfortably so that you need not be concerned about the body's safety or comfort. Take control of the breathing. Breathe in deeply through the nose and exhale through the mouth at least three times...

Now, with as vivid an imagination as you can muster, think a golden glow. Let energy rise within you as perhaps a feeling of joy. As you imagine this golden glow, and you feel that rising energy within you, the two meet in the neck area, forming a base across your shoulders.

The energy takes the form of a pyramid with a point at your chin and a point above your head. So you have a three sided pyramid with a base, the triangle of which is formed between each shoulder, from one shoulder to the other and to the chin.

Let the feeling of that pyramid encase your head, resting on your shoulders... As this golden glow surrounds your head in the shape of a pyramid, you begin to feel some kind of sensation of energy which moves up through your power center in your throat... on up behind your eyes... into the center of your brain.

If this becomes too powerful, dismiss it and do not continue until another time...

For those of you who feel the necessity to dismiss the pyramid, I suggest that you simply relax and enjoy the energies of those who continue with the exercise. You too will receive some benefit.

Now, those of you who plan to continue, become very aware of that golden, glowing pyramid around your head and the sensation within it. As the pressure builds, allow the pyramid to expand so that it absorbs the energy and does not cause you physical discomfort.

Then the pyramid becomes large enough to encompass your upper torso...

Allow it to rest, and begin to enjoy the energies that pass through you between the walls of the pyramid. You will find great stimulation, but also great healing...

Now these pyramids are also emitting this energy into the room so that those who are not participating may absorb the golden energy...

These energies passing through your upper body and head from wall to wall of the pyramid are stimulating your psychic centers.

Relax and allow any pictures that will to make themselves observable on the screen of your mind. Just enjoy the experience...

Now while you are in the golden pyramid, the pictures, the symbols that you see will bring you messages, some of them reincarnational, some of them from this life span. They will also inform you of your future.

As you perform this meditation, when you have the pyramid firmly formed and glowing brightly, you may ask whatever it is you wish to know from your own mind centers, and the walls of the pyramid reflect it back to you in the form of symbols which will answer your question.

This meditation is to open your psychic centers, to intensify your awareness. We have spoken before of the necessity for porousness. This meditation opens the psychic pores somewhat in the same nature as a steam bath would do for the body...

Now at first, you should not spend overlong in the pyramid, you should grow into it gradually.

You are now invited to put your golden pyramids in the center of the circle that the golden light shall flow together...

Invite the ones you love to come and join you that their awareness shall also be awakened. (Take time to speak names aloud)

Let the golden energies form a single pyramid in your midst. While you bask in its golden glow, and your beloved also, allow that pyramid to grow and grow and expand until it is large enough to completely encompass the globe that people everywhere will be touched and their awareness stimulated. Remember that your thought has the power to encircle the globe, not once, but a number of times before it loses its force.

So as you visualize this golden glowing pyramid, you are touching everyone wherever that pyramid exists. It has form under the Law of Illusion even in the realm of the invisible to the earth-physical. It is not invisible in other realms....

Now, with your personal blessing, allow your loved ones to return to themselves. Bring your consciousness back to this immediate room and to the increased awareness of yourself, your body, your circumstances, your environment, and all else around you.

Go in peace. And so it is done.
(Ref. S-33 Sheariam)

Four Sided Golden Pyramid

(Personal Guides and Healers)

When you have become sufficiently relaxed, place your palms together and generate the blue and green healing energies*. Visualize the blue and green as you breathe deeply...

When you have generated the energy, turn your palms toward your chest. Allow those healing energies to penetrate your own bodies, especially in the heart area. Feel yourself becoming warm and charged with this energy...

As that chakra becomes stimulated, the power of love begins to flow. It calls forth those of your own Band who are here to help you, to heal your own body, and to work in the healing of others...

When you are ready, bring your palms together again and rest your hands in your laps....

Now, behind each of you steps forward a spiritual guide. This is not your medical healer. This is a guide who teaches you in the matters of spirituality. And each places the hand above your crown chakra. Keep your palms together so that the energy that comes from your guide into your crown chakra is retained within the body...

As this energy comes into the crown chakra, you feel your head begin to fill with light... It moves on down to meet with your own energies in the heart chakra. And you feel uplifted.... Be

*See Glossary

silent for a moment while you enjoy this intimate contact with one of your own... In this silent communication, wounds of the heart are healed. Wounds of the spirit are bound up. Energies are refilled...

These guides step aside now and allow your personal healers to come near. You will feel the gentle heat of your healer's hands on the sides of your throat and neck. You may find your shoulders gently massaged. As that love flows through your body, it becomes very relaxed and receptive to the healing energies...

Your healers move to rest in front of you, placing their hands upon your feet. Your Spiritual Guide places a 4-sided golden pyramid over your shoulders, encasing your head. The energies that come in to you from the inner sides of the four walls of the pyramid are balancing energies. Their stabilizing energy permeates every cell of your body...

Each outer wall of your pyramid is like a transmitting station that transmits your thoughts and your feelings -- broadcasts. Deliver now the love, the healing energy through your thought transmission. Let that energy broadcast and fill the room so that all energies join together in the powerhouse of strength...

As these energies light up this room and join together, call forth those for whom you have concern that they too may benefit from this healing. (Take time to speak names aloud)

Rest now, and let the golden energies flood yourselves, flood the room and become broadcast throughout your entire globe. Just relax and let it happen...

Your healers rise and join hands forming a ring around all those who have come at your call. With this circuit of energy, they bless them all, each and every one, and guide them on their way back to themselves...

Now your monitors* remove your golden pyramids. Your own band* encircles you with their arms, giving you their blessing and assurance of their nearness at all times...

And so it is done. Go in peace.
(Ref. S-34 Sheariam)

---

*See Glossary

Baptism

(Pool of Life's Waters)

When you are ready, begin to control your breathing. Breathe slowly and rhythmically, in through your nose and out through your mouth.

As you breathe out, visualize your breath joining the breath of all others as though a pool of life energy is forming in your midst.

In unison, call upon Khamael*. As you repeat the name three times, feel those waters of life becoming heated with the Divine Fire.

"Khamael, Khamael, Khamael."

Now just feel the warmth as your feet rest in the pool of life's waters. As the heat warms your feet, it moves up your legs and into your torso and out your arms. And your whole body becomes warm and relaxed.....

Now invoke the strength of Gabriel* and feel the strength of that power moving through the waters of life. "Gabriel, Gabriel, Gabriel."

Now as the strength moves through the waters and follows the pathway of the warmth, your body becomes strengthened with eternal strength. Your heart beats firmly, slowly, steadily, fulfilling its function with strength and ardor. All the functions of your body respond to this strength.

---

*See Glossary

As your body becomes relaxed, and warmed, and strengthened, your own healers come to you. Healers from the Temple also join you. Feel them as they stand near you. Feel their hands as they move upon the parts of your body that are in need....

As you each breathe out, continue to visualize your breath as the life waters, filling the pool in your midst as it becomes like a baptismal font... If you will completely relax, become very loose in your body, your own healers with those from the Temple will take the subtle astral body and baptize it in the fountain.

Thus you feel washed with the waters of life, cleansed of your transgressions, cleansed of your worries, cleansed of your apprehensions, of your doubts, your fears -- all those things that produce discomfort in the physical organism are washed away in the baptismal font of the warm living waters...

When you are ready, ask your healer to remain in the fount and baptize those loved ones you call in. When you have received the consent of your healer, call the names of those you would have served.....

Now within yourselves, feel the cleanliness. Feel the burdens lifted. Feel the release of the baptism. And as you and your loved ones rejoice in this healing, sing the "Hallelujah" before finishing the meditation.

With joy in your hearts, and knowing that you have been blessed, allow your loved ones to return to themselves. And in your visualization, sprinkle the globe with these blessed

tion, sprinkle the globe with these blessed life's waters, and know that the healing service goes on and on to all who would accept. With your spirit you baptize the globe on which you live. And so you create an environment in which the Law of Harmony can be enacted. And so it is done. Go in peace.
(Ref. S-31 Sheariam)

A White Pyramid On A Desert Road

(Blessing by a Spiritual Teacher)

If you are ready, settle yourselves comfortably so that you can take your attention away from the physical body.

Imagine, if you will, that you are standing on a road in a desert-like panorama. The sun is shining, but it is close to the horizon. Imagine also that the road stretches infinitely behind you. And it stretches infinitely before you.

As you walk along the road, you are following the sun, so it never sets upon you. As you do this, become aware that time is no longer a particular for you. Time is as limitless as the infinite road behind and the infinite road ahead which allows you to stay always in the sunlight.

So on this road you have all the time you need or could ever want. Thus you find yourself freed of one of the confinements of which you are usually conscious.

Now as you walk forward on the road in the sunlight, become aware that coming up over the horizon are the tips of pyramids. Because you are not limited by time, you may move as quickly or as slowly toward those pyramids as you feel comfortable doing.

One of the pyramids is sparkling white in the sunlight... Then another appears to be golden, radiant like the sun... As you move, other pyramids come into view. You see that they are of different colors, all beautiful in the sunlight...

You are invited to come into the white pyramid. As you approach the base of the pyramid you will see the door... There you will enter a narrow passageway -- dry, cool, welcoming, comfortable. Not too far along, this narrow passageway opens into a reception room comfortably furnished...

You are invited to find a seat and make yourself at home. Look around you. Enjoy the appointments, the furnishings, whatever is there that draws your attention...

Now from an opening in a wall in this room, enters a Being robed in white. This Being moves near you and blesses you. As you receive the blessing, you experience an expansive feeling as though suddenly you could understand all the mysteries of the universe...

As you experience the interchange of energy between yourself and this Being, you know that what you learn is for everyone... Ask permission of the Being to invite those whom you would to join you. Receive the being's gracious acknowledgment of your request and permission...

Invite those whom you would to join you in the white pyramid... (Take time to speak the names aloud). Be not concerned whether your invitation is accepted or not. Those who will learn, those who wish to share, will join you. And those who are not ready will choose not to do so...

As your guests enter, they also receive a blessing, and you feel very united, all of you as one great thought... In this union, relationships are healed. Little irritations are set aside. Even great differences become unimportant...

Relax. Bask in the glory and the glow of this Being. The words you hear are like music to your soul. You find you are at peace within yourself... Radiant with vitality.

Now the Being rises, blesses all of you, bids you return to your daily lives, knowing that you will take with you the radiance which you have now experienced. So you part, moving out of the pyramid by way of the passage through which you entered... Return to the road.

Even as you bring your consciousness back to daily routine, the road awaits you. Return to it often. Seek out that which lies within the other pyramids. Go in peace, and in love.
(Ref. S-41 Sheariam)

Three Sided Golden Pyramid

(Exploring New Regions)

Relax your bodies. Breathe deeply. Focus your attention within yourself. Become aware of your subtle bodies. Allow yourselves to feel an expansion...

Again work with the golden pyramid. Visualize the 3-sided pyramid -- the flat side across your back coming to a point in front of your face...

Now as you become aware of your expanding subtle bodies, you become aware that the pyramid seems to pulsate with light... Allow that pulsating to become quite warm... The golden pyramid acts for you as a transcendent energy, and you feel your consciousness rising and expanding with your subtle bodies.

As you transcend the physical body, you move into a rarefied atmosphere where you become aware of knowledge, of love, of peace, of truth.... As you contact these eternal principles, you become infilled. The force of truth brings healing into the subtle bodies. The presence of peace establishes security...

As peace and truth permeate your subtle bodies, your physical bodies respond. The energy systems of your physical organism align themselves in proper order...

With the golden pyramid you have the freedom to explore regions which you have not previously attempted. Allow yourselves to expand, to move higher and higher. Enjoy the freedom...

Now, from this expanded stance, from this sense of freedom, it is as though you can gather the healing energies in your arms like textured clouds. Gather in these energies. Bring them into the midst of the circle.

These pure vital rays of energy are all around you... As you concentrate them through the focus of your consciousness, the room lights up, and there is healing for any who would come. So allow a portion of your consciousness to speak through the physical organism and invite those about whom you have concern. (Take time to speak names aloud)...

Enjoy for a moment more the healing energies available to you now. As you are transcended beyond the physical organism, you can be much more sponge-like. There is all the healing energy necessary to serve you and all your loved ones, and plenty left over...

Still with your golden pyramids in place, draw back into your physical organisms. Feel again throughout your body the pulsating light of the pyramid. As you allow your loved ones to return to themselves, taking with them the healing which they choose to take, be conscious that with each pulsation in your own being you are dispersing this healing energy you have gathered. You are wafting it around the globe making it available to all who would accept, all who choose to reach out and accept the healing they desire...

Let the pyramid grow more and more transparent until its light dissolves into the ethers, awaiting the time that you shall collect those energies again into the golden pyramid...

As the light disperses, and your loving thoughts go out, close your meditation with a song.
(Ref. S-36 Sheariam)

Golden Pyramid

(Extend Your Lovership)

Gather your thoughts for a meditation. Put yourselves at ease. Focus your attention in the center of your midst, and visualize there the golden pyramid...

Allow that visualization to become vibrant. Let the pyramid grow so that its etherealness fills the room. Even though you are inside of it, you are also outside of it.

Place your hands upon the walls, palms flat. Feel the golden light go into the palms of your hands. Let that infusion of goldenness move throughout your entire body...

Let that goldenness represent for you the highest order of Divine love... That goldenness pervades your body, so that every cell, every corpuscle is beloved... The goldenness becomes the Divine Lover... Let those beloved cells, every molecule of your being, love you...

Now your hands glow as though they are incandescent light. Place those hands on your body wherever you have an impairment you wish to heal. Let the Divine Love make you whole...

If, in your perception, your hand dims, place it once more upon the pyramid wall -- there is sufficient healing and love and light -- you may have all you need... Take it in until your body glows...

As you become filled with this Divine Glow, there is that within you that must share it. Extend an invitation to all about whom you have concern.....

Welcome all who come into this pyramid of light, of love and Divine Healing... Accept this blessing for yourselves and for your loved ones who asked to share...

And now, as this golden light, this Divine Love circulates among all of you, visualize that golden pyramid becoming larger and larger until it indeed encompasses your globe. Extend your lovership to all leaders of all nations, to all those who would ask for help wherever they may be, in whatever condition they may find themselves...

And thus you bless your planet, and you give of yourselves to right the consciousness that there shall be a future, for your civilization to grow, to become, to fulfill its promise.

With love, release those who have come in answer to your invitation. Let them return to their own consciousnesses. With great love from the Temple, we bless you and keep you in our hearts. We bid you go in great joy and peace. And so it is done.
(Ref. S-77 Sheariam)

Herbal Treatment

(With Personal Healers)

Begin to breathe deeply and relax your bodies. Feel your healer standing behind you with hands upon your shoulders. The warmth of the healer's hands moves through your body...

Now the healers invite the water bearers, and they create a communal pool into which the warm water is poured. You are invited to put your feet into the pool. There are aromatic herbs and oils and essences added to the water.

As you feel the energy from your healer's hands move through your body, you should feel the warmth in your feet and the relaxation of the herbal treatment. Even as you breathe in, you inhale the aromatic emanations of the water... As these vapors fill your lungs, the healing moves through all the systems of your body.

The healer's hands direct the healing action... Breathe deeply drawing in these healing vapors. These will assist your body in combating the problem that is abroad in your environment at this time. (Pollution, for instance)

As you relax beneath your healer's hands, you will find your scalp massaged and your temples gently stroked. Your spirits fill with an ecstasy...

Enjoy the ministering hands of your healer's touch. Let your consciousness rise. Illuminations of love and joy permeate your being.

While your physical body remains calm, cosy and comforted, your subtler bodies become stimulated, and you feel an expansion. As these subtle bodies expand, you unite one with the other. Begin to sense the oneness which you really are..

Now as you expand into each other, feel the love and joy of each and every other person flowing into yourself, and yours flowing into them. Experience the genuine union as though you were all fingers on the same hand...

Now your healer places hands again on your shoulders, and you are free to call your loved ones to come stand in these warm healing herbal waters. (Take time to speak names aloud)

Let the joy within express as a child. Allow yourselves to get completely into the pool. Splash water on each other, frolic, and enjoy..

Now your healer awaits you with a towel to dry you off. You are patted dry... Return to yourselves. Those whom you have called are permitted to return to themselves...

The water bearers come to take the pool, lifting it from your midst. It is taken into the air, and its waters mingled with the waters of the clouds. Thus the healing waters are carried around the globe to be deposited with the natural rain...

Do not forget to give thanks to your healer for the service. Express your gratitude always when you are served. And so once again it is done.

Take with you the joy and the healing and the love which you have shared. Bear it with you always. Know that you are always and forever linked with the others. Go in blessedness. (Ref. S-35 Sheariam)

Synchronizing Rhythms

(Sounds Are Felt)

Settle yourselves comfortably in your chairs. Turn your attention inward. Become aware of the breath going in and out of your lungs. Be aware that you can control the muscular action which draws breath in and which expels it.

Witness the breathing in and breathing out process. Become aware of the rhythm of life. Then if you still your body very still, you become aware of the pulse beat of your heart...

Through the awareness of the rhythm of your breathing and the different rhythm of your heart, and by becoming very, very still, you find the awarenesss of the life pulsation throughout the tissues of your body. Each cell, each organ functions in its own rhythm according to its own nature.

When you are very, very, very still, you become aware of the flow of the blood through your veins; that eternal circulation which is a tiny replica of the circulation of universal energy throughout the body of expression which encompasses us all.

Now allow your attention to move outward through the external portion of your physical organism. Become aware of the subtle vibratory rhythm of your auric field. As you develop this sensitivity, you become aware of the organization of energy which surrounds your physical organism, and you are aware of the movement and the rhythm of that organization.

As you tune in to this finer vibration, your physical organism may even feel as though it is blocking. That is its response to the organized rhythm of the subtle bodies which surrounds the physical organism.

Allow your awareness to continue to expand until you become aware of the rhythm of the people who sit on either side of you... Then as you become aware of those varying rhythms, you find your rhythms changing themselves into a synchronization so that the subtle bodies begin to throb in unison. As this occurs the group truly becomes one pulse beat.

Now in this unity of rhythm, feel within your emotional body that sense of peace and wholeness which represents perfect health. As you move into that consciousness, the tissues of your physical organism respond and you experience healing.

This energy is present and available for all. The Invisibles about each of you, and the healers from the Temple, join also their energy and bring the organization of perfected wholeness, so that each in his own way can tune into that united vibration and carry the healing with him as he leaves the group. You may also, by voicing aloud the names of those about whom you have concern, invite their consciousnesses to join in this vibration.

Continue the expansion while your loved ones come to join you.

Allow this synchronized vibration of peace and wholeness to move out from this room, moving throughout your community, throughout your nation, and around the world, that all who are of like mind may join and synchronize their energy and thus attune the planet's consciousness to the polarity of peace.

Even as the carrier waves of your thoughts perform this service, you enjoy a sense of upliftment, of lightness, and a gratification that you once again have contributed to the positive polarity and moved mankind forward, no matter how infinitesimally.

Give thanks to the presence of the Invisibles who have come to join you from rarer realms. Release your loved ones to return to themselves, taking with them the healing and peace. Go in blessedness.
(Ref. R-6 Sheariam)

Beauty Treatment

(Healing Lower Extremeties)

Breathe deeply. Load up your systems with oxygen... (*Request the company and services of the Temple healers, then quietly await them.)

Relax your bodies, and allow the healers to administer to you. As they massage your feet and encourage the cells in the feet to absorb more of the pure oxygen, you will feel your feet become very warm, even tingling.

Each will be served in his own way according to individual needs... Some will have small bone adjustments made. Some will have energy put into the connective tissue. Some will have tension withdrawn from the muscles. All will receive a treatment of relaxation. Allow it to happen. Flow with the ministers' hands... Enjoy this delightful ministry.

As the healers work on your feet, and bring the bones into proper alignment, enabling the muscles and the connective tissue to do their job properly, so are adjustments being made in the organ of your understanding...

Feel the circulation increase as the healers move their treatment on up your ankles, your calves, and your knees. Those of you who need special treatment on the knees may feel the urge to straighten your legs. This is caused by the healers pulling your legs straight in order to work on that joint.

---

*See footnote Page 25; see Appendix

Continue to breathe deeply to give the healers the oxygen necessary for their work....

As you know the healers' treatments flow in accord with the lessons being given, and thus you will receive a massage of the thighs also. (Refer to the Principle of Migration and Differentiation*). The thighs represent in the body your ability to move about on your own, upright, with strength. Thus you get the treatment now on your knees and your thighs...

Already, as the healers work on you, the blue and green energy begins to collect in the room. Feel it about you. Let it add to your healing process....

Now as the healers complete their treatment, they move outside your circle and join their energies one with the other as they stand behind you. You can feel their energy coming through your back. You may feel the warmth of their person as they stand behind you.

The blue and green energies swirl around. The Healers are spinning them into beautiful shell-like conformations... I am told this is a treatment of beauty; that each in his own way is having that which is beautiful within him, charged up, emphasized, expanded. And then this energy over-glows into your lives and that which you have around you also becomes more beautiful.

Into this room filled with life giving energy, healing thoughts, and expanding beauty, call those loved ones about whom you have concern. (Take time to speak names aloud)

As you breathe in this beautiful energy, your body becomes filled with its beauty and its healing. As your loved ones come, their spiritual bodies, their subtle bodies also become filled with this beautiful healing energy. Soak in it as in a warm bath.

Now help the healers as they swirl this energy into whirling vortices and send it out throughout the world. Visualize this with them. Add your power... The energy then is transmitted out of this room into the surrounding community, and indeed throughout your globe. All is uplifted by the beauty; made whole through the healing.

So you have added your efforts that all shall come to completion in the appointed place at the appointed time... A this energy is sent forth, the celestial music begins. Listen for it...

Now allow your loved ones to return to themselves, and together with the healers raise your voices in song. Bring your meditation to a close with the "Hallelujah".
(Ref. S-28 - Sheariam)

Invisible Healing Hands

Settle your bodies. Focus your attention on the palms of your hands. Feel centers of energy like pools of hot liquid in the cups of your palms...

As though your palms do indeed hold a hot liquid, allow that heat to permeate the flesh of your hands, to move up your arms, and into your chest area...

From there let is spread throughout the rest of your body until your body is thoroughly warm through and through...

Once you have become thoroughly warm, direct that warmth to any portion of the body which needs to be re-aligned, re-attuned, or reaffirmed... As you focus your attention upon that portion of the body, you will feel it become very warm....

While you heal yourselves from within, the Temple healers come close, placing their hands upon your invisible bodies in those areas which pertain to that portion of your ailing. Thus a response is set up, a pulsation as you feel the heat from within meet the heat from the healers hands...

Now to complete a circuit, the healers place the other hand above your head. You will feel a drawing from that portion of your ailing body up through your head. The healer's hand draws forth the discomfort, the symptom. As that is withdrawn, the other hand assists in the reorganization, the re-attunement, the reaffirmation...

In the next few moments the healers will be releasing, bringing their hands together above your heads, pouring out into you their blessing and love. Receive the healing with a willing heart. Accept the blessing. Express your gratitude and appreciation to your healer...

Now again, let the warmth congregate in the palms of your hands. All those from the Temple do the same; that warmth becomes a blessing of healing.

In a process similar to evaporation, allow that warmth to move up from everyone to form a cloud-like formation in the atmosphere. Put your love and your blessings into that cloud-like formation.

Now the healers work with those who have chosen to come. Bless them and give thanks...

The healers bless them and allow them to return from whence they came. Join the healers, directing with your thoughts the cloud formation to surround your planet, to make available to all who would that same blessing of healing, love and peacefulness...

Return the focus of your attention to yourselves in this room. Know that the thoughts of peace and healing continue to bless those who will accept. The healers and other members of the Temple, moment by moment, continue to reinforce the work that you do.... Go in peace. (Ref. S-63 Sheariam)

Circulation of Healing Power

(Blessing & Sanctifying The Body)

Relax yourselves and focus your attention upon your feet. Feel the circulation of the blood as it flows through your veins. Since mind controls matter, the focusing of that attention brings a warming sensation to your feet. One can always understand better when one's feet are warm....

Now let that warmth move up your legs, increasing the circulation in your knee joints and in your hip joints, so that your physical organism becomes more flexible...

Let that warmth continue to rise, flowing out through your shoulder joints, to your elbows, to your wrists and warming your hands...

The Temple healers approach you*. Place your palms facing each other apart a few inches. Feel the power of the life-giving force moving from the hands of the healer into your physical organism, increasing the power of the field between your palms...

Now, bring your palms together, closing the circuit, locking this power within the physical organism to heal, to comfort, to make whole. This is the blessing and the sanctification of your physical organism....

Feel now this power rise to the center of your forehead becoming very focalized. As that

---

* See footnote page 25

focalization becomes more than you can contain, allow it to pass from you, melding with all the others in the center of your circle...

You continue to be healed even as you create a field of healing life force in your midst into which you may invite anyone for whom you have concern. (Speak the names aloud.)

Hold your hands palms upward in a welcome to those who respond. Welcome them with love.

Be thankful for the healing that you have experienced, and the healing that is available to them....

When you are ready, rest your hands comfortably in any position that you wish, but continue to allow that life-giving power that is being channeled through you from the healers and from yourself to gather in the center of your circle...

Ask those who have come to join you to work with you, to send this life-giving force around your planet, to bless those who are in peril, who are in terror, who are in pain, or who have any need whatsoever. Visualize energy extending to all who have need around the globe...

Release those who have accepted your invitation to go their own path feeling enriched. Go you, now, in blessedness and peace.
(Ref. S-65 Judith)

Reflecting The Absolute

(Actualizing the
Experience of The Absolute)

Relax your abdomens by breathing deeply. Allow any tension in your neck and shoulders to flow away from you. To stimulate the healing energy, place your hands approximately two or three inches apart, palms facing, and move them in a rotational manner, building up an energy between your palms. If you cannot feel the energy, begin by placing your palms directly upon each other and rubbing them briskly. Then allow them to part gradually.

Now breathe gently into that ball of energy forming between your hands. As you exhale your breath, realize that that is part of your own energy system. You are transmitting your emotions of love. Your thoughts of healing go with your breath into that ball of energy between your hands.

This is not the universal energy. This ball of energy is actually comprised of some of your own auric field. By giving this field structure with your hands, you have caused some of your auric field to become condensed.

You can now turn your palms upward and blow gently upon them, pushing that ball of energy into the center of your circle where all the balls of energy join together into one great field of potentiality.

There it spins and sparkles and glows. As you continue to transmit your thoughts of healing, your feelings of love into that ball of energy, it grows greater and greater. It gains in potentiality. As it grows, it comes back to touch you at your toes, at your knees, at your hands, and at your forehead...

As this energy touches you, you become balanced and attuned with each other in this room. Accept this balancing. Accept the attunement.

You can feel this as the sides of your body come into attunement with each other, as your back feels straighter, and your head sits more easily on your shoulders. Accept this healing energy into the internal organs of your body, so that all of the functions within your body become balanced and attuned.

Since circulation is important, you not only accept this balance and attunement for yourself, you also return it to the center by your thought, by your love. The more you take in, the more you give back, the greater the potential. This is a miniature description of the act of creativity on the part of The Absolute as It sits in Self-contemplation and circulates Itself through Itself, gaining greater and greater potential, actualizing Itself in multitudinous ways. As you participate in this circulation, you are actualizing the experience of The Absolute.

The potential grows greater and greater, becoming strong enough to have more than needed by those present. By announcing the names of those for whom you have concern, you offer invitation for them to join you here in their consciousness.... As the invitations are accepted and your loved ones respond, the healing field gains a great potential. Even the Invisibles come close to share the action.

Expand your awareness. Become aware of those who have joined you in the invisible. Some of you are even now being touched. Accept the love, accept the healing, and return it. Those who have answered your invitation now partake of the healing potential. Transmit your thought of blessing to those who have come.

In this expanded awareness, you may also approach any member of your group and offer a blessing. As you bless one another, you establish ties which will persist across distance and time whether you are together or not.

Lift your hearts in thanks to the realms of the Invisibles, to the angelic realms who have come to join you in this effort. Know that their thoughts of peace are being poured into this field of potential, that there shall be healing throughout your planet.

As this field of potential is fed, so it grows enormous and moves away from this room, growing and growing so that it surrounds the entire planet. It carries your love, your thoughts of peace empowered by the invisible realms.

In this powerful moment, make that affirmation "LET THERE BE PEACE" (repeat three times.) "LET THERE BE UNITY IN THE HEARTS OF MEN" (repeat three times). "LET THERE BE LOVE" (repeat three times).

And so you strike the key notes. Accept your blessing. Feel within yourselves the gratitude, the appreciation for the cooperation you receive from on high. Establish within yourself the consciousness that PEACE IS... UNITY IS.... LOVE IS. And where there is peace and unity and love, there is joy.

Release your loved ones to return to their own consciousnesses, to return to their own affairs. So you are released to your own affairs. Go in peace.
(Ref. R-9 Sheariam)

Your Star

(A "Homing" Device)

Settle yourselves. Relax your minds. Take deep breaths and relax your bodies...

Allow your mind to imagine you are separate from your body, free in time and space. Fly away from the atmosphere of this planet. Free yourself from this solar system. Be free in space, as though you yourself were a planet or a star...

Allow yourself some concept of limitlessness, weightlessness... but also that complete integrity of being — that completeness, that wholeness which is your Self...

Then look around you until you see a star which appeals to you. Know that that star is an entity, even as you are... For this time, acknowledge that star as being a part of your own higher consciousness, which can be for you a homing device toward which you can turn at any time you may become lost.

Because that star is an entity even as you are, know that you can communicate, that you can experience a relationship, an interchange of consciousness... To that star, communicate a goal, a dearest wish, an objective, a desire...

Now, reach out your hand to the star, and ask for a stream of its light to place itself in your hand.... Feel the connection of love...

Still clinging to that beam of light, bring your consciousness back to this material body upon this planet... Know that at any time, you may look to that star to be reminded of your objectives, to find the method by which your goals can be fulfilled. The star does not do it for you. It is a homing device, a navigational point.

Keeping contact with that part of your higher consciousness, stays you more truly upon your own pathway. You can at any time feel that beam of light within your hand, which reminds you that higher guidance is always available to you...

Now look to your star; see it glow very brightly. Know that that is a blessing of illumination to your planet through you. You bless the planet upon which you live by virtue of the fact that you exist upon it -- do not forget that fact...

As you go about your daily lives, radiate that knowledge, that brilliance of your star, and thus illuminate all of your planet... Know that you are blessed. Live your life in harmony, and travel your path in peace.
(Ref. S-62 Sheariam)

A Walk In The Meadow

(Rejuvenation and healing)

Raise the level of your consciousness by joining in this meditation. Rejuvenate your spirit; bring yourself healing.

With the powers of creative visualization, walk through a meadow. Note the shades provided by certain stands of trees. Feel the sunshine that warms your head. Note the beautiful color in the flowers that spring among the grasses of the earth.

As you walk along together, you begin to hear the celestial music that plays all the time. It is only in moments of quiet relaxation, when you take your attention away from the dross world, that you can begin to hear the melodious chimes.

As the music surrounds you, you are warmed by the sun and cooled by the shade. Your feet are welcomed by the springing grass in the soft earth.

Be aware that you are joined by those whom you love. Turn to them and call them by name and welcome them on this walk with you. Those you don't see, invite.

Cross the meadow and walk up that little rise over there. Find comfortable seats on the ground beneath the trees. Your spirits are high and joyous as you are bathed in the warmth of the sun. You are cooled by the shade of the trees. The melody surrounds you. The vibrations of dear mother earth rise from beneath you, and the joy of the companionship of loved ones permeates your very souls.

When you feel love bursting from within so intense as to be almost unbearable, you want to share this marvelous joy. Reach out and touch each other. Bless each other as love flows through and around you, from you, into you. For a moment in silence, enjoy....

Now as you shared these ecstatic moments of love, joy and companionship, you realized that you are whole. You no longer have bodies of infirmity. You no longer have limitation. You are totally free, yet held securely in love.

Look about you. See that others are healed. They, too, are whole, joyous and beloved. The warm sunshine is like liquid gold, and the earth holds you in its mother arms.

Even in this holding and sharing, know that you cannot clutch or grasp or retain. Release. Let go. You are free as you free everyone else, yet you are linked with ties of love. You are filled with a serene calmness. There are no cares, only joys and peace. You come to a common understanding. You are all of one mind, yet free -- uniquely individual.

In that knowledge of that individuality, again bless everyone around you. Begin to walk down the slope and across the meadow. Bring with you the joy, love, the calmness and the serenity you found on that little hill. The celestial music still rings in your ears. So you come back to your point of departure, having shared a moment of healing, that joyous love. With a final blessing, enclose yourselves again in your individual lives, knowing still that you are forever linked....

Now come back to the consciousness of this room. Bless each and every one of you. God speed. Go in love and peace.
(Ref. T-12 Silverthistle)

A Visit To The Temple

(Imagination and The Astral Realm)

Put yourselves into a comfortable position. You are invited to take a trip. In your imagination, see yourself on a rather high plateau. Mist surrounds you. You have a sense of having left your material world...

Feel your body becoming light... Enjoy a sensation of freedom... Now, as you stand on that high plateau, become aware of all the members of your group there with you, and notice that they seem to be ephemeral, ghost-like. Know that you are seeing them in their subtle bodies...

Now notice that in one direction there is brighter light than in the other direction... Gaze gently in the direction of the light and you will see my form coming toward you...

As you function in this imaginary plane, you are entering what you commonly think of as the astral plane...

Now follow me as we move towards the Temple... There is not much scenery on this plane, for it is a mental plane, but if you wish scenery you may create that which pleases you.

We approach the Temple. See it as it glistens and glows. Come with me through the door and we will move into the entryway and turn into the corridor to our right. We will go a short distance along the corridor. Then I invite you to enter the library. As you go into the Library you will see walls and walls and stacks and stacks of books and parchments and rolls and scrolls on every topic you can possibly imagine. Browse a while....

Now I invite you to come with me to the healing-meditation room. We shall leave the library and go further down the corridor as it follows the circumference of the outer wall. Here we find the door of the healing room. It opens for us and we enter. Within the room you find comfortable seats, places to lie down if you wish.

You will notice that not only the Temple healers are present but the healers from your own spirit guide bands. All the healers are present and they begin to serve you. As they pass their hands over your body, your blood becomes purified, the toxins withdrawn.... Any pockets of noxious energy within the body are extracted and purified... Spinal adjustments are made where necessary. Any other alignment necessary is made.

Each of you receives a gentle scalp massage while another massages your hands and another massages your feet. For a moment or two just relax and enjoy this...

As you relax enjoying your massage, you become aware of a fragrant mist permeating the air, and you feel your sinuses and your lungs cleared out, relieved, purified. Now the healer who has been massaging your scalp moves gentle fingers down to the tender tissues around your eyes, and indeed gives you a treatment of the entire facial structure...

As needed, each massaging healer works on those parts of your body which need relaxation, toning up......

Entering the room now is one of the beautiful high priestesses of healing. She radiates healing light and the room fills with her light as she smiles upon us all..........

And now the joy, because of the healing light, leaps higher in our breast, and we must share. So you may call any beloved to join you in the healing room.

.......And now she blesses us again, and bids us depart from her presence. She does not leave us, since we would feel in the dark as she takes her light with her. So as soon as you are prepared to leave her presence, leave the healing room and move back to the plateau. Absorb as much of her light as you can before you leave..

She will bless all your loved ones whom you have called and who have responded, and send them back to themselves in her own good time...

Meet with me now on the plateau that I may bless you.... I bid you go in peace and harmony. Blessings to you all. And now I shall leave you. Go back to your material body.. You are refreshed and healed, and it is done.
(Ref. S-31 Siverthistle)

The Opened Cube

(or THE CROSS)

(Just prior to giving us the following meditation, Silverthistle gave us a homily about the salt crystal. Since the meditation refers to his words in the homily, the pertinent reference is included here.

"In your mind draw a picture of a square. Then draw a square on the top side, a square on either side of the original square, and two more squares in line at the bottom of the original square. This is the cube opened up, a symbol of the cross.")

Settle yourselves and build the energy. By rubbing your palms together, light up the room as though there were neon throughout the atmosphere. It is beautiful. Relax now and bring in your loved ones. (Take time to speak loved ones' names.)

Now again, build the energy, and as you direct it to these loved ones which you have called in, see it forming a cube-like structure enclosing them. Visualize this very strongly...

After the pattern of the open cube which has just been given, allow the sides of light to lie down on the floor and become the symbol of the cross upon which your loved ones are standing. This is a pathway of healing light for them, and the emanations continue to move up through their bodies permeating them with healing.

The cross stretches its arms out. The energy moves into you and through you, and around you and brings healing to you as well.

Now let the Christ, represented by the Cross, rise and overflow, carrying the light throughout the world. You become the healers, the spiritual man embodied in earth, that you are supposed to be. Put lots of power into your visualization. See the light emanating from the cross, moving throughout the world to bless and to heal, and especially to bring peace to those tortured souls who have so recently died by violence.

This is the true missionary work of the Christian church. And so it is done. Bring your hands up, palms to your chest. Feel the light and the power. Let this blossom within you, blessing your lives so you bless the lives of all with whom you come in contact. Now placing your palms together, you complete the act, and you are blessed. And so it is.
(Ref. S-14 Silverthistle)

Loving The Body

(A Ceremony of Joy)

Settle yourselves. Clear your minds. Breathe deeply.

Using your imagination, move away from your body to the center of the room. Then turn to face your body as it sits in the chair. Observe it as it sits there, and feel yourself as separate from the body.

Note first the familiar things about the body... Then take note of anything which seems unfamiliar to you...

Now gently, until you become accustomed to the exercise, allow yourself to look at the other bodies sitting in chairs around the room. Note the familiar... Take note of the unfamiliar, that which you had not previously noticed...

Look back at your own body, and experience within yourselves a love for that body, even as you might look at your own child and know that you love that child. Look at your body and love that body...

With affection look at other bodies in the group. Move around from one to the other. In each case, love that physical manifestation...

Even though your consciousness is outside looking at the body, the body's sensory system accepts the warmth of the love being directed towards it... Not being hampered by the consciousness, the cells of the body can bask in that love and be healed...

Now while the bodies rest comfortably in the chairs, let your non-physical selves mingle in the center. Hug each other, love each other...

As you exchange these thoughts and these emotions, you build a powerful field of energy, a tremendously magnetic field of loving thought and emotion. It is like the contagion of joy. You cannot help but feel uplifted.

As you participate in this ceremony of joy, think of those loved ones that you would invite to join you. (Speak their names aloud.) As they receive your message, they leave their physical body resting at peace, and their non-physical parts join you in the group. Let there be great loving, embracing, and sharing....

Now, using the power of your imagination, think of quantities of this love energy as balloons. Like children, toss those balloons into the air, and let the currents in the sky carry them around the world...

So you share this joyous energy you have brought together. Fling your arms. Bat those pieces of energy around. Push them around the world with your mind... Enjoy this. Have fun. Be child-like.....

Now with tenderness and with sweetness, embrace and kiss goodbye. Allow your loved ones to return to themselves. Each of you return to your physical body, and feel the enlivening action in the physical cells as this joyous, uplifted, happy, non-physical self returns to the physical body. Feel the stimulation in the blood stream. Feel the lifted heart in the breast.

And last of all feel your connection with each other. Know that as you share love and joy, you heal and are healed. You are blessed. Go in peace.
(Ref. S-50 Sheariam)

Lilies Of The Field

(Feeling secure brings joy)

Imagine, if you will, being seated at the very heart of a lily. See the beautiful petals rising round you, wafting perfume for your enjoyment, shielding you from the elements of wind and rain, but open at the top so that the warmth of the sun keeps you cozy.

Then peek out from your nest between the petals that surround you, and see all the other lilies of the field in which are nestled others. Know that while you are cozy, warm and protected, you are not alone.

Now, as the lily draws sustenance through its roots, it brings that energy up through its stem, and you are able to absorb it into every cell of your body. You are nourished. With that gracious nourishment, the very consciousness of each cell sings with pure joy. You know that you are whole. You are perfect. You are complete.

From this safe, secure, nurturing place, let that singing joy rise from you. The more it rises out of you, the more there is to rise. And the air is filled with joyous healing sounds. You breathe it in deeply and circulate that joyous sound in and out.

The more you breathe in and the more you breathe out, the more joyous sound issues forth. It becomes a hymn of peace. In this way you broadcast peace. Any and all who seek for peace of mind, who seek for health of body, may breathe in that joyous sound and be restored.

Now with the power of your higher spiritual self, direct that joyous, living, healing sound to anyone for whom you have concern. If that person is ready to accept peace, accept healing, accept wholeness, accept perfection, it is all there for him.

You broadcast it freely, lovingly. You are thrilled by the very act. Know that this joyous sound is propelled throughout the world of your planet by the power of your loving thought. It becomes available to anyone everywhere. It rises as a paeon of praise to Divinity Itself. The realms of the angels and the archangels join in the holy song. You are lifted to dwell with them for a short time...

But that rarefied atmosphere does not support you well. You must return to your own realm. As you do, the joyous sound remains in your heart, remains in the very consciousness of your cells. Your body sings with health and balance, purity and wholeness. You go forth from this place in blessedness.

Go with God, in peace and blessing.
(Ref. T-13 Silverthistle)

Garden Of Eden

(Eradicating negatives)

Make your bodies comfortable. Allow your breathing to be regular, normal, and comfortable......

I invite you on a trip of imagination to the Garden of Eden. Visualize it in any way that you wish, as it can only be visualized in accordance with your belief system....

See yourself standing in a beautifully landscaped area, landscaped by the very Spirit of God. The magnificent trees soar above your heads. The flowers and the shrubs offer their blossoms and their scent for your enjoyment. The bees happily buzz from blossom to blossom. The birds sing their songs. The sun shines and the night creatures sleep as the day creatures sun themselves...

Now, although you are primarily day creatures, in this visualization you are creatures who are awake both day and night. So allow the sunlight to fade and the darkness of night to envelop you in a radiant warmth.... As the day creatures go to sleep, the night creatures awaken to enjoy the sounds and the smells and the light of night...

Then, again the light returns. The night creatures sleep; the day creatures awaken. In both conditions of day and night, you were awake, alert, aware, and enjoying it. Allow these days and nights to revolve around you for a moment.

Be aware that you are without fear, without shame, without sorrow, without pain. You are whole and perfect as an original creation. Your environment is completely friendly....

Now, in one of your daytime periods, decide or choose something in your life that you wish to explore. Find yourself a place in the Garden to settle down, either to sit or lie, whatever posture is comfortable for you.

Then in your mind begin your exploration as you allow the night to envelop you. As the sunlight fades and the night envelopes you, become aware of the sense of freedom that the darkness permits you...

This darkness is the recess of your innermost mind. Thus it can bring to you the results of your exploration. Knowing that you are protected and cared for and loved, allow the night to bring you whatever it will...

When you are ready, allow the sun again to lighten your environment. Bring into the sunlight the knowledge of the night, the experience of the night, the emotions and the feelings...

Nearby you, you will find a pond of crystal clear water in which you may bathe. You may wash away any fears which remain. You may cleanse any of the thoughts which may have arisen in your exploration. You can heal any wounded emotion. Allow yourself the luxury of baptism in the pond...

When you are through with your self-baptism, allow yourself to dry in the warm air. Invite your loved ones to join you for their own self-baptism, if they so wish....

Now the Temple people join you, and all together, you enjoy the peace, the tranquility, the perfection, and the sharing....

We bless those loved ones who have come, and everyone is released to return to his or her own consciousness in the daily world in which you must function. Bring back to this daily consciousness your experience, your feelings, your healings, and your sense of being cleansed. Go in peace.
(Ref. S-69 Judith)

The Amethyst Light

(A protection against radiation*)

(Ask for the help of a Temple healer and while you quietly await the contact, complete the following relaxation instructions.)

Relax those muscles which hold your head erect, which tighten across your shoulders, and down your mid-back... For you to benefit to the greatest extent from this very special treatment, it is important that your shoulders and neck be relaxed.

Allow your head to sit gently upon the column of your neck in line with your spine.

When you are approached by a healer, you will feel a slight pressure on the very mid-top of your head. The healer will open the channel through your crown chakra to allow an inpouring of some very special Cosmic force. You can assist in this process by visualizing a column of amethyst light coming from some reaches above and entering that mid-point of your head, and channeling down your spinal column...

The amethyst light carries with it a defense mechanism similar to your antibiotics, similar to the action of the iodine, that will assist your physical organism to resist the effect of radiation. Allow the amethyst light to move throughout your spinal column, fully, effectively, until you feel a warmth at the base of your spine as

---

*See Appendix

the light stimulates and impregnates the kundalini*.

When you have had a sufficiency, the healer will close the crown chakra so that you may retain within the protection of that amethyst light. If you wish, you may then invite your loved ones to join you. Those who do, will also be served. (Take time to speak the names aloud.)...

Now, even as your loved ones are being served, those in realms above bring down to you a great shaft of golden light which arrives as an inverted pyramid...

That light is your defense from the outside as the amethyst light is your defense from the inside of your physical organism. As often as you othink of it, take from that shaft of golden light and surround yourself with it. Automatically those who are in your immediate vicinity will also benefit from that protection.

You cannot, however, take that light and deliver it to anyone else. You may only appropriate for yoruselves and your immediate vicinity. That light is a permanent gift to you (who have asked) and to those loved ones who have answered your invitation...

Move into that golden light and dance and sing with great joy of spirit...

Use the memory of this meditation as a rpompter for you as you go your own way. Release your loved ones to return to themselves. Accept the blessing of the healers. Beloved, go in peace. (S-74-Judith)

*See Glossary

Spirit Of The Tree

(Communion With Nature)

(Prepare for this meditation by preparing the golden bubble as given in the previous meditation)

Settle yourselves to be at peace...see in the side of that bubble a sliding door through which you may step. Move out of the bubble, and out of the building in your visualization. Go out among the trees which surround you.

Find a tree to your liking... Approach the tree gently — remember the tree also has subtle bodies. Approach gently and place your hands lovingly on the bark of the tree. Feel the life flow -- that tree is a living entity, as you are...

Expand your awareness to make contact with the spirit of the tree. Tell the tree that you love it, and feel the tree return your love... When you are sad or lonely, down-hearted or blue, you may go into nature, make contact with some natural thing and receive comfort, because it also lives.

Because these natural things live, they benefit by your appreciation and by your love. You cannot make this intimate contact with a living, natural thing without experiencing a change in your own consciousness. Spend a moment with your tree as you would spend a moment with a very dear friend...

The magic of this connection is that that feeling of community can be taken with you even though you leave the presence of the tree. Tell your tree "goodbye" and wish it well, and return to the golden bubble.

Seat yourself within the bubble, and commune with yourself for a moment about the healing you have experienced through the contact with the tree.....

Now allow the large bubble to return to the individual bubbles so that you are encased within your own golden glow. Keep the glow around you. Radiate peace. Share the golden light throughout your remaining time together. You will find great enrichment which you will take back to your ordinary day, and carry that blessing with you for ever. Blessings.
(Ref. S-60 Silverthistle)

Fountain Of Blue And Green

(The Spectrum of Healing)

If you will breathe deeply it will relax your body. Rub your palms together briskly and generate some energy. While you are rubbing your palms together begin to visualize blue and green light....

Now open your palms toward the center of the circle and begin to allow the blue and green light to accumulate in your midst... See it shimmering like a huge ball of irridescent energy... The blue and green are the spectrum of healing.

As you put these energies into the center, allow them to form in your imagination into a fountain ..... as though there is a fountain of blue and green light rising out of the floor.... Visualize this fountain of blue and green light enlarging to a greater circumference... The energy then comes back to shower you with healing.

Accept it. Let it permeate you. Let it flow into you.... and back out of you.... even as the life force is granted you.... As these energies pour in upon you, you will feel refreshed. You will feel relaxed...

Get your visualization as strongly as you can. See the fountain your midst. Feel the blue and green energies as they are sprinkled on you and absorbed by you, then transmitted back to the center again...

Now you have a wonderful opportunity to share this healing experience with those for whom you have concern. If you speak their names aloud, and they so choose, their consciousness will be drawn into your circle....

Continue to visualize the fountain. See now those loved ones as they come in to share in the healing energy of the fountain...

All the while you continue to be blessed by the healing energy, visualize the fountain growing higher and higher, greater and greater, taking this healing energy, these thoughts of peace out into the city around you, out into the surrounding terrain, clear across your country and indeed ultimately surrounding your world...

In unison chant "peace... peace... peace... So you set a great wave in motion. Know that it goes out to touch everyone, and there will be those who will respond. Thus you have given blessing and in turn you are blessed.

Release your loved ones to return to themselves. Know that your thought of peace continues washing upon the shores of the consciousnesses of all it encounters. Carry within yourself the consciousness of peace, healing and blessedness. And so you are blessed. Go in peace.
(Ref. R-4 Sheariam)

Pink Healing Cloud Of Love

(Peace Affirmation)

Settle yourselves for a healing meditation. As members of the Temple gather round, and the healers come forth to pass among you, relax your bodies. For these next few moments, clear your minds even of that which has been spoken here this night. Let yourselves be receptive. Open your hearts knowing that the room in which you sit is sanctified, is cleansed. It is blessed, and has been prepared especially for you, to share in this moment of communion and healing...

Take some deep breaths and relax your abdomens. If you would be healed, relax your bodies. Many of your ailments and complaints stem from tension, far more than from disorganization of your energy fields...

As the healers gather round you, they lift their hands upon you, and radiate their love out to you... In your mind, think of that light as a pink cloud, sparkling with pure light behind it as you might see a cloud at the sunset. Know that that pink cloud contains the healing power of love.

As it comes to you, it goes through you and onward to your neighbor, and through your neighbor, and on... Say the word PEACE in unison three times. Draw it out so that the energy goes forth from your heart. Let love be behind your words.

P E E E E E E A C E ..., P E E E E E A C E ...,
P E E E E E E A C E ...

If you have a specific request in your mind, allow the healer to touch your body.... As you feel that warmth and healing, give a thought of thanks. Bless the healer who has come to you. Then release that which you have claimed as an infirmity. Let it go. Disclaim it. Claim your healing, and affirm it to yourself. Be blessed in that healing.

Now, if there are those of your loved ones about whom you have concern, invite them to come in consciousness to join this meditation.... (Take time to speak names aloud.)

In united effort, direct that cloud of energy provided by the healers so that it shall spread through the atmosphere, making itself available wherever there is pain and suffering. Wherever there is a need, that need can be met. You have yet to realize the powers of your mind. You can, with that which you now know, distribute this energy of peace and love throughout your planet...

Now release your loved ones to return to themselves... Bless all those who have come to join you from the Temple. Go in peace, in love, and in holiness. And so it is.
(Ref. S-46 Karansa)

Breathe Pink

(Cosmic Love)

Relax your bodies. Be as comfortable as you can. Breathe in through your nose as deeply as you can and exhale through your mouth. Let this inhalation and this exhalation be gentle but deep. Be aware that you are appropriating cosmic substance with every intake of breath.

Now give some thought to the biological operation that is going on within the body as you breathe in a lung-full of oxygen. It enters the bloodstream and collects the toxic waste. As you breathe out, you are expelling those waste substances. This biological function is a necessary function for the continuance of your planetary life. What is toxic waste to your body becomes the nutrient to other bodies. So as you breathe in energy, you breathe out energy.

Now let us call this energy "love". Be aware that as you breathe in, you are appropriating cosmic love. The oxygen goes through your blood stream, carrying cosmic love to every part of your body. When it has deposited the cosmic love, it picks up the negative energies, the toxins and the wastes, which are expelled with your exhalation to become cosmic love for some other form of life right here on your own planet.

Just think about that for a moment or two as you continue to breathe in the cosmic substance of love. Feel that cosmic substance of love moving clear down into the tips of your toes, tingling the tips of your fingers, filling all parts of your body with love.

A nice rosy pink hue is the color that goes with this cosmic substance of love. As you expel your breath, breathe out pink. Take in the cosmic substance of love and breathe out pink.

Pink is also the color of magnetic attraction. It draws to it all the other pink. As you breathe out this pink, it coheres into a beautiful bundle of rosy tinted energy.

With every breath you expel you are adding to that rosy energy. It becomes larger... growing and growing until the room is filled with it. You can begin to feel that soft magnetic touch on your skin.... Still you breathe in the cosmic substance of love from the limitless supply.

There is not only enough for you, but as you breathe it out, it becomes enough for all else on the planet.

Your thoughts and your emotions and your desires have created a beautiful healing environment of lovely pink. Invite into it by name those for whom you have concern. Let them bask in this pink energy.

Even as you bask in the rays of this beautiful pink, and your beloved ones are healed in its energy, every breath increases the volume. It begins to brighten up not only this room, but also the surrounding environment. It moves out of the house and into the street, and subtly surrounds everyone. As they breathe in and breathe out, even unconsciously, they add to the vibrancy. The rosy glow grows greater and more intense and more healing.

It begins to encircle the globe like a beautiful sunset. There will be some who will be able to see it and marvel. There will be those who have the consciousness to know what it is and will appropriate what they need for themselves.

There will be those who will pass through unaware. But like the evening breeze that cools the cheek, they will feel something, and they will be better for it.

Now although you will be leaving this room shortly, you will continue to breathe. Every time you become aware of the breathing process, you will become aware simultaneously that you are breathing in the cosmic substance of love. Your body is being cleansed.

With every breath you breathe out, you are breathing nutrient and healing to other people. You will find this to be a blessing each time you think about it.

Now let your loved ones return -- free them from your thoughts. Send them with loving blessings, back to themselves. Even as you end this meditation, be aware that the energy field is still vibrant, still active, that you are surrounded by the rosy energy. Feel its uplifting, its cleansing action, and be joyous in it.

Now as you close your meditation, bring the palms of your hands together to lock in the energy, to close the circuit. You are blessed, and it is done.
(Ref. S-21 Sheariam)

Rose Energy

(Love)

Breathe in and out gently. Observe your breath. Now as you think about your breathing process, think about breathing in life and breathing out love. Thus it is that breath should be expelled gently. You must be gentle in your loving.

Become aware of your body, of your mind, of your heart, as a circulation point for life to come in and be released as love. As you breathe in the life, your body becomes energized and you feel the energy rising. As you gently breathe out love, the energy that goes forth from your body is in a rosy hue. It has some red of the life energy, it has some pure whiteness of the spiritual energy, and it has a tinge of blue of the mental energy. This delicious rosy hue emanates from you as you release your breath. It permeates the room. Let each feel the glow.

Even now, the atmosphere within this circle begins to glow like a sunset. Just as the light of the sunset covers the surface of the earth, so does the energy of love go forth from this circle. It lights up the horizon for any who would lift his eyes to see.

As you gently release your breath, your love goes forth, this beautiful wave of rose-colored energy. It flows throughout the earth. It offers itself to the bodies of those who need healing.

Each of you now may bring into the circle to be bathed in this rosy glow, any for whom you have concern.

Then each of you, in your own way, take from this pool of rosy energy all of the healing you would have for yourself. Even now as you breathe in, you breathe back in this loving, loving energy. As life and love are cycled and recycled, they grow ever more powerful. The cells of your body are regenerated. The hearts of your loved ones are blessed.

Now just let that rosy energy find its way around the globe. Release it -- Let it go.

Now clasp within your own forms the healing you have claimed for yourselves. Each are blessed. So you know you are beloved. And it is done.
(Ref. S-19 Unidentified)

The Purple Cloud

(Fragrance of Heaven)

Relax your bodies and minds. Detach yourselves from your daily concerns. Unite yourselves as one with the affirmation spoken aloud in unison:

"I AM A CHILD OF GOD"... "I BLESS MY BODY"... "I BLESS MY THOUGHTS"... "I BLESS MY FEELINGS"... "I AM A CHILD OF GOD"...

Begin your meditation by rubbing your palms together... Bring forth those beautiful purple tones from your heart... Speak the names of the loved ones you wish to invite...

Now again rub your palms together. Turn your palms to the center of your circle. Keep the gorgeous purple light going. It is so much a higher vibration than the blue and green. Let your loved ones bathe in that blessed healing energy.

The purple cloud is so vital and so moving, it spreads and envelopes you and permeates you, bringing healing to the consciousness of every cell of your body and the bodies of your loved ones. It carries a lovely fragrance, the very fragrance of heaven itself...

With your powers of visualization spread it throughout the world as a beautiful purple cloud enveloping the planet, spreading its fragrance through the streets of every town and every city, and along every country road til all who would participate and accept the healing are healed and blessed...

Open yourselves to the healing. Soak it in through your pores. Feel it move throughout your being and return it into the world. As it goes around and around the planet, it takes not only healing but comfort. Even those who are not ready to be healed are comforted....

And so it is done. Bring your palms together, locking the healing within yourself. Know that you are blessed.
(Ref. T-7 Silverthistle)

The Purple Flame

(Angelic Help)

Relax your bodies. Breathe deeply to relax your abdomen...

Visualize at the pit of your stomach a cone-shaped structure of energy which moves outward from you in concentric circles... expanding energy... moving out from your center... like a small whirlwind.

As you feel this tug in your solar-plexus, become familiar with that tug. It is one of the signals the Auphonim* uses to express their presence.

As that energy moves outward from you, and the tug becomes stronger, you begin to understand the necessity for the Malachim* which apply their energy from the other direction, going the other way. So there is a balance. You are familiar with the opposing forces and their action upon each other...

Now as this whirling vortex of energy emits from your body, visualize at its very center point a deep, deep purple like a tiny flame... Then allow that purple to spread out along the rings of your energy vortex, becoming lighter and lighter as it becomes more distended, until by the time it reaches the center of the room it is nearly purely white...

*See Glossary

This energy moves in your circle, joining with the energies of others, forming a structure of pure white energy. The Malachim provide the environmental restraint to hold this energy within the circle.

As that accumulation of clear white energy becomes sufficient, you will feel the tug at your center diminish until there is just a very calm, peaceful glow in your midriff...

Allow that glow to permeate your entire body. Just let it spread...

It feels cool, comforting, and uplifting as it moves through your body. You feel your body becoming more balanced in all its parts...

Feel the calmness in your heart area...

In this calm, you can express peaceful concern by speaking the names of those about whom you have concern. You can invite their consciousnesses to join you by walking into the white energy...

If there is within you any place in your body or your mind or your heart which does not seem to be at peace, you can also partake of this white energy. This white energy is like water -- you can put your feet into it, or your hands, or your whole body. You may drink of it. Treat it as you would water, healing water...

Visualize that purple flame receding into your innermost parts where it shall continue to burn as you go about your daily affairs. You can call it forth at any time you wish*.

Fold within yourself the healing which you have appropriated. Bless those who have come to share with you. Allow them to retreat to themselves.

Allow this energy which is like water to flow along the surface of your planet. Let it be an invisible cleansing flood, bringing cleansing power to all on Earth that it touches. But, inasmuch as it is the white energy, it will do no harm as that much water would do. Those whom it touches may partake of it as they will, or let it go on by.

The Auphonim and the Malachim will take care of its distribution, its availability. You have served your fellow man by providing the physical aspect with which the angels may work.

In a moment of silence, give thanks to the angels, to your teachers, your guides, your ministers. And bless yourselves as you go in peace.
(Ref. S-55 Sheariam)

---

*If you choose to do so at home alone outside of a group, be sure to call upon the archangels before you begin the exercise.

Color Spectrum

(General Blessing)

This time we are going to work with different colors. As you generate the energy, first see red, a physical, highly energized red that contains no hostility.

In this way, this clear, bright red stimulates the consciousness of the cells which comprise the physical body. Now as you visualize this red energy, allow some blue to come in, bringing it into pink and lavender. Build the good, strong field of energy of the pink and the lavender.

These colors raise the consciousness of the cells which have been energized by the red. As you breathe deeply, you make the blue become more brilliant, and the lavender turns to a bright purple. Feel the warmth and the vibrancy of the richness of this glorious color you have generated. Let it swirl around you as though a mist. Now, into this beautiful mist cloud of purple, call your loved ones for mental and emotional and spiritual healing as well as physical. (Take time to speak the names.)

As you see your loved ones in this circle enveloped in this cloud of purple energy, direct your thoughts to it, and see it slowly begin to change into shades of green. See it iridesce, and then move on into the spectrum of yellow, and it gets very bright and becomes very golden. This very bright yellow spectrum is the color on which thought waves move throughout the universe.

So now see these energies moving out away from you, bringing the healing and blessing thought to all, everywhere. As it encircles the globe, it comes back to you. As the yellow approaches your red physical light, it becomes orange, and you are refreshed and stimulated as well as healed.

Now bless your loved ones and let them depart. When you are ready, bring your hands together, closing the healing energy to yourselves. End with a personal blessing for yourself alone. And so it is done. You are blessed. (Ref. S-16 Silverthistle)

Prosperity

(Invocation of Universal Consciousness)

Breathe deeply to relax your bodies...
Tune to Universal Consciousness by chanting the AUM 3 times ("Ah-oo-mm, Ah-oo-mm, Ah-oo-mm").. Now let that connection become solid for you. Be at peace...

Now, with that increased rate of vibration, call upon Khamael 3 times ("Khamael, Khamael, Khamael")... Feel the warmth of the Divine Flame which burns within you as life... Affirm "I am eternal" 3 times ("I am eternal. I am eternal. I am eternal."). Feel that reach...

Now, call upon Gabriel 3 times ("Gabriel, Gabriel, Gabriel."). Feel the solid strength activated within you... Affirm "I will to be strong" 3 times ("I will to be strong. I will to be strong. I will to be strong."). Feel the power...

Now, call upon the name of Haniel, 3 times ("Haniel, Haniel, Haniel"). Feel the supply.... Affirm, "I have overflowing abundance", 3 times ("I have overflowing abundance. I have overflowing abundance. I have overflowing abundance."). Know that completed, filled feeling, and then more...

Even as you are filled with love, with prosperity, with health, with self-expression, with happiness, it overflows and becomes a flood of all abundance. Invite now your loved ones to join in this beautiful energy. (Take time to speak the names aloud)...

Celebrate this infilling, this blessing by singing the hymn of thanks, "Hallelujah"...
Now allow your loved ones to return to themselves. Allow this overflowing blessing to spread out from this room, move out through your community and indeed around your planet...

As you bring your meditation to a close, be sure to bring your palms together to lock the blessing inside. Go in love and peace.
(Ref. S-37 Sheariam)

Invocation Of Archangels

(Meditation for the hostage situation
Beirut Airport, June 1985)

You will be calling upon Khamael, Gabriel, and Tzadkiel. Your third affirmation will be "mercy and love are infinitely reproduced".

So as soon as you are ready, settle your bodies. Calm your minds. Set aside all concerns of your own, and concentrate. Focus your minds on that landing area where your compatriots sit in such grave situation.

Breathe deeply. Relax your body so it does not conern you. Relax your mind so that as you call upon the archangels, the orders of angels can carry forth the intention of healing.

Relax your emotions so that you do not feed in hostility, anger, or resentment, only compassion and concern. **But not anxiety.** Calm, peacefulness, and certainty that you are having effect...

Say the Aum (AH-OO-MM) three times to focus all of your energies together. ("Ah-oo-mm, Ah-oo-mm, Ah-oo-mm.")

Then in unison: "Khamael," three times ("Khamael, Khamael, Khamael."). Then affirm 3 times, "I am Eternal Life. I am Eternal Life. I am Eternal Life.")... (pause) "Gabriel," three times ("Gabriel, Gabriel, Gabriel."). "I will Divine Will. I will Divine Will. I will Divine

Will.")... (pause) "Tzadkiel," three times ("Tzadkiel, Tzadkiel, Tzadkiel."). "Mercy and love are infinitely reproduced. Mercy and love are infinitely reproduced. Mercy and love are infinitely reproduced."

Now, as these centers are awakened within you and the orders of angels come at your call, rub your hands together and generate physical energy. Toss it in a ball into the center of the room... See the blues and the greens, and the pinks, and the lavenders... Be at peace. Be peace.... Feel compassion...

As the healers from the Temple join you and commit their energies, invite also those whom you will of like mind to join you in their consciousness to further this effort... (Speak the names aloud.)

Continue with your mind and heart to build the energy so that at the proper time the angels of Tzadkiel can carry this energy of peace, of mercy, and love to the proper places... Feel that yearning within yourselves that the physical safety, and the emotional safety of all concerned is accomplished. Feel that love and compassion and yearning for a world at peace in which you may live in peace...

Know that you are calling upon the highest powers available to you to accomplish this mission... "Mercy and love are infinitely reproduced." Reaffirm that to yourselves...

As you lend your minds and your hearts, and with the Temple healers and teachers, see the power grow. Visualize the colors. As Tzadkiel sends forth the angels, they will carry not only the energy but also the colors to affect at the intuitive level those who have the power to change the situation, to change the balance of power to a safe resolution...

As mercy and love are infinitely reproduced, sanity is brought to an insane situation... Visualize the angels in any form that makes sense to you. See them carrying that energy, those colors. Move them directly to that landing field, to that airplane, into the offices where heads of state make their decisions, where those who hold the power are holding their discussions.

Let that energy and those colors attach to communiques that understanding shall be reached between the parties... Let this energy of mercy and love clear the very air breathed by these participants...

Allow now those whom you have invited to return to themselves, but do not draw the meditation to a close. Carry it with you in your hearts as you leave this place. Keep it in the forefront of your consciousness until this situation is resolved. Know that each time you remember this meditation activity, that you are enhancing and empowering. Continue the affirmation as often as you can... "Mercy and love are infinitely reproduced."

As you take your separate ways, you may say your farewells by singing "Let there be peace". (Ref. S-52 Sheariam)

Invoking Divine Fire

(Courage, Strength, Power)

When you are ready, take some deep breaths and concentrate for a moment or two on relaxing your bodies. Breathe in lots of oxygen so that all parts of your body are energized...

Make the connection with the Divine Energy. Recite the AUM 3 times together ("Ah-oo-mm, Ah-oo-mm, Ah-oo-mm")... Allow that connection to become affirmed...

Now together, repeat the name of Khamael 3 times ("Khamael, Khamael, Khamael.")... In the silence feel the power of the Divine Fire that enlivens every cell of your body. That is the power behind every cell of your thoughts. Feel the fire burning within you.

So you begin to find that the power rises from the base of your spine, and slowly moves upward... By repeating the name of Khamael, you have invoked the very Divine Fire within your own person, within your own entity.

Again together, repeat the name Gabriel 3 times ("Gabriel, Gabriel, Gabriel.")... Feel the strength as the rising Fire within reaches the chest area, and you find courage, strength, and power. Your will is strengthened, and you realize that you can do and be anything you want...

Now with fullness of the heart let that love center expand. Feel the warmth of the Fire. Let the healing pervade your being... Transmit the Fire to the center of the circle...

With your thoughts spread this Divine Fire throughout your globe to carry courage, strength and righteous power to those who seek healing for themselves and for your planet. Clasp within your own bodies the courage, strength and power you have gained. Go in peace. So let it be.
(Ref. S-29 Sheariam)

Expansion And Dissemination Of Invoked Power

Begin by relaxing. Breathe deeply...

Follow the breathing with three Aum's. ("AH-oo-mm, AH-oo-mm, AH-oo-mm.)... (pause)

Invoke the power of Khamael. "Khamael, Khamael, Khamael. I am Eternal Life" (pause)...

Invoke the power of Gabriel. "Gabriel, Gabriel, Gabriel. I will Divine Will." (pause)...

Invoke the power of Tzadkiel using the affirmation "Mercy and love are infinitely reproduced".

"Tzadkiel, Tzadkiel, Tzadkiel. Mercy and love are infinitely reproduced."... (pause)

As you feel the influx of the power, allow that power to build into great force. Feel it heal your bodies... You are Heaven and Earth. Deliver this message to the cells of your body; to the cells of your auric bodies.

As the power pours in, expand yourselves. Feel yourselves becoming one... When you are united as one, there can be no separativeness, no inequality...

As you expand, give invitation to those for whom you have concern. (Speak the names aloud.) Include those in charge of your hostage situation. (Even if you do not know the names.) Invite them to come in consciousness to share in this unity, in this power of healing....

As the members of the Temple join you, as those whom you invite come in their consciousness, the power field expands.

Reaffirm "Mercy and love are infinitely reproduced. Mercy and love are infinitely reproduced. Mercy and love are infinitely reproduced." Feel that affirmation in the cells of your being... Feel its movement in your psyche. Know that as you are moved by mercy and love; so are those whom you have invited and who have come....

Now, with the power of your mind, with the thought that travels round your globe, spread that mercy and love to the atmosphere. Make it available to all who would partake of it. It is tangible, measurable substance. As a mover, move that substance throughout your planet. Make healing available...

Finally, as you have been taught, seal your own healing within your own bodies, (by placing the hands together and drawing them in) .... Express your gratitude to your Source, to the members of the Temple, and to the Divine Beingness of which we are all a part... Go in continued peace and serenity.
(Ref. S-53 Sheariam)

Ed.Note: This meditation followed the one on page 153 by two weeks. Immediately following the first meditation came the dramatic change in the hostages' circumstances.

Healing For Emphysema

(Healing Invocation)

Relax your bodies. Relax your heads. Let your hands rest comfortably on your laps. Think about your body. Mentally review your body. Wherever you find tension, dismiss it.

Take deep breaths into your nose and out through your mouth. With every exhalation allow the tension to pass from your body. Allow this tension energy to aggregate in your midst into a big ball of vibrant tension energy.

Expel all the tension energy from your body to make a ball in the center of your midst, strong, vital, powerful. The more you relax your body, the more powerful the ball of tension energy in the center becomes.

Now this tension energy is extremely dehydrating to the body.

When the water retention is very much out of balance, as it is when one suffers from emphysema, there is fluid in the lung tissue. Picture to yourself a pair of lungs. If you cannot visualize that, think of two pink sponges rather oval in shape side by side. These two pink sponges reside in the chest cavity.

Mentally direct the dehydrating ball of tension energy to move through the solar plexus, gently, gently. Softly and gently it enters the solar plexus. You can feel it.

It moves upward into the chest cavity. It begins to dry those overburdened cells. Dehydrating action removes the excess fluid from those pink sponges. It's a gentle action, soothing. Now affirm wholeness (for yourself or for the one whom you are treating.)

Then let the energy begin to dissipate, taking with it the excess fluid, leaving the pink sponges whole, perfect, and clear.

As the energy leaves, taking the excess fluid, it restores balance in the body fluids. So there is a balancing action both inwardly and outwardly. Wholeness becomes perfection.

Invoke the Law of I Am. "I am whole. I am whole. I am whole." Now let that energy rest within you.

Invoke the Law of I Will. "I will heal. I will heal. I will heal." And let the energy rest within you.

Now as the power of Khamael and Gabriel is dispatched into your midst, call the names of those for whom you have concern. Let them too be blessed. (Take time to speak the names.)

Affirm for yourselves and for all you have called in, "We are whole in the power of Khamael. We are whole in the power of Khamael. We are whole in the power of Khamael". Let the energy rest.

Affirm "We will heal in the strength of Gabriel. "We will heal in the strength of Gabriel. We will heal in the strength of Gabriel". And let the energy rest.

Now rest in these powers which are aspects of The Absolute. Allow those whom you have called in to return to themselves, taking the healing with them to be sustained by Khamael and Gabriel.

Before you close, affirm "I am peace in Khamael. I am peace in Khamael. I am peace in Khamael." And let the energy go forth from the group throughout the world.

Affirm "I will Peace in Gabriel. I will peace in Gabriel. I will peace in Gabriel". And let the energy go forth into the world.

And so all are blessed. Go in peace.
(Ref. S-22, Sheariam)

Pillar Of Divine Fire

(Invocation)

Visualize in your midst a pillar of light. Make it as big around as you wish; make it as small as you wish; but make it a bright shining pillar of Divine Fire.

As this pillar of fire glows mightily, feel the radiative energy coming into you. Then together say "Khamael" three times. Follow this with the affirmation "Disciplined forcefulness is mine" repeated three times.

> ("Khamael, Khamael, Khamael. Disciplined forcefulness is mine. Disciplined forcefulness is mine. Disciplined forcefulness is mine.")

Now just feel that energy. It's as though the pillar of Divine Fire moves into the interior of yourself. You feel it there lighting your body; glowing into all of your subtle bodies.... This gives you the power of self control -- self-determination.

Again affirm: "I am disciplined forcefulness." Just feel how that makes that pillar glow hotter.... until your whole being radiates with light.

Say "Gabriel" three times followed by the affirmation "Eternal Beingness is mine" repeated three times.

> ("Gabriel, Gabriel, Gabriel. Eternal Beingness is mine. Eternal Beingness is mine. Eternal Beingness is mine.")

Now feel that radiant energy. Feel the Spirit Truth flow throughout your physical body; flow throughout your desire body; flow throughout your emotional body; flow throughout your mental body.

Again re-affirm "I live forever".

Just feel that permanence, that indestructibility that is truthfully your Selfs.

In the pillar of light, let that glow expand and expand. Call in your loved ones to share the blessing of the Divine Fire.

Now, by the power of Khamael and by the strength of Gabriel, you are all blessed with the healing energy of Eternal Perfection. Let the pillar of light grow larger and larger.... Then it becomes a beacon of truth throughout the world. All who would accept its Light are healed. All are one.

As you visualize the Pillar of Light growing and growing, the light fills your room, and indeed surrounds the entire globe. Allow your loved ones to return to themselves, taking the blessing of healing with them. Feel yourselves blessed and healed.

Let the light go on — the Eternal Flame burning always within yourselves. Thus as you go forth from here, you are also beacons of truth. Blessed are you. And so it is done.
(Ref. S-24 - Sheariam)

Chanting "Love"

(The Freedom Love Gives)

Let us begin with three deep breaths. Focus your attention at the center of the crown of your head. At that focal point, draw all the tension up your body to that point. Feel it coming up out of your feet and legs, your hands and arms, all the way up through your body, and wrap it into a tight ball of tension at the crown of your head. As you draw this tension out of your body, all of your muscles relax and your body is at ease.

Bless that ball of tension energy, and let it flower and blossom and spread away from you above your head. As you bless it, it becomes the energy of love instead of the energy of tension, and it illuminates the entire room. To the psychic eye you all look like candles...

Release that energy and turn the focus of your attention into your heart area. Become aware of the beating of your heart... slow, rhythmical... dependable.

Together, chant the word "love" three times similar to saying the "Aum"...

("Ll-uu-vv, Ll-uu-vv, Ll-uu-vv.")

Feel the expansion in your chest. Feel the freedom that that loving energy gives you.

Now go into the center of that expanded area, into the place of "I AM". Feel yourself at the very core of this expanded love energy. Revel in the warmth, in the delight, in the joy of being surrounded by pure love...

As this feeling grows, you feel that you must let it out. Again chant "love" and let it expand. Let it fill this entire room...

("Ll-uu-vv, Ll-uu-vv, Ll-uu-vv.")

Enjoy the blessing of the love which has expanded to you from others as yours has expanded to them. Breathe it in... Soak it in...

In this highly elevated consciousness of love, in this light-filled room, there is the healing power. Call those whom you love... (Speak the names aloud.)

Now again chant "love"... Let the energy permeate your loved ones, yourselves, and overflow this room, sending the energy out around the world...

Celebrate the healing with the "Alleluia" (group sings). Now rest in the beneficent healing energy for a moment or two... Again feel that warmth of love as a kernel within your breast. Place your palms on your breast; then bring the palms together to lock in the healing energy.

And so it is done. You are greatly blessed. (Ref. T-9 Silverthistle)

The Fireball

(Influencing Those in Power)

Build the energy. Now, take your palms slightly apart and feel the energy as a ball resting between your palms. Roll that ball around until it becomes very large. Then throw that ball into the center of the room.

By this time your awarenesses should be expanded to the degree that you can feel that energy as all the balls of energy meet together in the center of the room and form a whirling vortex of energy. Visualize your blue and green lights...

That whirling ball of energy begins to throw out sparks... The blue and green light grows brighter... Indeed the room lights up. Now, into this beautifully lit room call your loved ones for healing. (Take time to speak loved ones' names)

The energy is very high... Let it permeate you; let it heal you.

Now, like a huge fireball, let that energy grow, grow and spread out, moving out into the countryside, moving out into the towns and into the cities, flowing across the surface of the entire globe, bringing healing and comfort to all it touches. Even as you visualize, send your blessing with it to give it added strength.

Remember that your thoughts have the power to travel around your globe a number of times. As you visualize this healing energy going forth, be sure that your thoughts are thoughts of blessing. Know that they reach afar from you.

Bless also the leaders of nations. Bless those that have the power to negotiate peace.

Give also a mental blessing to all those you have named this night -- increase their potential for receiving healing.

Now also, each of you bless the others in this circle. Then give yourselves a special blessing for your very own self.

And so it is done. You have built great power this night. You are very blessed.
(Ref. S-15 Silverthistle)

The Flame Of Eternity

(Contentment, Confidence, Joy)

Arrange your bodies... Breathe deeply three times.... Say together the AUM 3 times ("Ah-oo-mm, Ah-oo-mm, Ah-oo-mm.")... Let the Divine Energy permeate all the cells of your body. Be gentle and relax...

As you move now into the deeper silences within, feel the Divine Fire begin to warm the vital parts of your body. Visualize a Flame of Eternal Fire burning at the center of your being. As you visualize the Flame of Eternity burning within you, allow the flame to grow larger and larger. Feel the heat extending its warm healing energy throughout all of your body, into your hands and into your feet.

As you feel this spreading healing warmth, be aware of the peace within that requires no understanding... a deep abiding contentment and confidence in the Divine...

As you enjoy this spreading warmth and its contented peacefulness, feel joy arising like bubbles in your blood stream. As it rises, let it move out of your body into the center of the room. Let all that warmth and peace and joy co-mingle until it forms a flame in the center of the circle. Enjoy the brilliance, the warmth and the love.

In this glorious attunement call in your loved ones to join in the healing and to find peace and joy. Now as the flame grows higher and higher, it serves as a beacon. It calls into your presence to be with you many of the joyous souls that are about. Their energies are added to the group's, and they take from that fire and spread it throughout the globe...

Know that even as the flame is allowed to dissipate and the energy to encircle the Globe, you have within you your own individual flame of eternal life, of peace, of joy, and it shall be with you always. You not only warm yourself at your own inner flame, but you also warm those whose own flame has grown dim....

As your loved ones return to themselves taking with them the healing, close your meditation with that precious wish for peace on Earth. Sing the joy that is in your hearts.
(Ref. S-30 Sheariam)

A Dynamo Of Electric Blue Light

(Begin this meditation by speaking your name aloud.)

Relax your bodies; relax your minds. Breathe deeply, and be at peace...

Each in his own way, begin to visualize a rod of blue, electric blue, light rising in the center of your circle... Use your imagination. Use all of your powers of visualization. Make the rod of light as real to yourselves as you can... Just a simple column of electric blue light. Smooth the edges. Make it cylindrical... Empower it — make it stronger....

Allow it to spread out over the floor, as though it were a column of water. Let it carpet the floor, running under your feet and under your chairs and out behind you. Do not let it get holes in it. Keep it thick, keep it brilliant. Let it be very smooth...... Now you may relax.

Workers from the Temple pick up the edges as though this were a brilliant blue cloth, and pull it up behind and around you, closing it over your heads as though you are all enclosed within the petals of an electric blue flower. Begin to feel the security, the closeness, the intimacy, the intermingling of your energies as they are held within the ball of this closed flower...

Now in this very intimate circle, each bless each member of the circle. Expand your auras. Take into yourself all the others. Become one...

This brings together the energies which you brought into the circle with the expression of your name as a mantra when we first began.

You are linking your physical energies into a very powerful dynamo. (You are assisted in this effort because you have invoked, because you have asked. As we have said before, we cannot do these things for you; we can only help you do them.)..

Now you have built up a very powerful hot arena of energy. The Temple workers now pull back the enclosure, and you feel less confined, but still intimately linked with each other. As the Temple workers place the blue material back upon the floor, with your visualization powers, draw it once again into the rod standing in your midst...

The Temple healers come to assist you, and with their powers help you to build a very large, a very powerful rod which rotates. As it whirls about, it emanates a healing energy. Accept that healing energy into your bodies. You should be feeling strengthened, empowered, enabled...

As you take the energy into your own energy, you are balanced and harmonized.. Invite your loved ones who would to join you now while this energy circulates in the room... (Speak the names aloud if you would have the consciousness respond.)

Feel yourselves bathed in the energy. Your invited guests also bathe themselves in this energy...

Now, release your loved ones to return to themselves. With the help of the Temple workers, visualize the rod of energy rising straight up into the atmosphere, still whirling. As it whirls it begins to flatten and spread out throughout the atmosphere, whirling its way around your planet, making itself available to all for whatever purpose they need -- for their healing, for their change, for their transformation, for their understanding, for their harmony......

See the energy disappear into the atmosphere. Give a word of thanks to the Temple workers and the Temple healers. As you go your separate ways, you now carry a part of each other with you. When you work upon a common project, this link will serve as a dynamo to accomplish your goal. Go in peace with love. (Ref. S-56 Sheariam)

Rhythm Of The Auric Egg

(Green Light and Golden Ray)

Get yourselves settled comfortably. Become aware of the egg-like aura around each of you. Feel yourself nested within that egg... Even as you become aware of that egg, you realize that it interpenetrates the one next to you, and indeed flows throughout the room...

Become aware that those energies are in a pulsating rhythm. There are varying rhythms within the egg -- some are fast and some are very slow, but they all work together in a symphony of color and sound...

Visualize strongly a very clear, green light emanating from your solar plexus, as a ray goes forth from the sun...

As the ray from each of you meets in the center of the room, it forms a pool, a reservoir of healing energy. To the degree that you can feel the rhythm in your auras, that healing power of the green ray is enhanced....

Now visualize a very clear golden ray coming forth from your heart and blending with the green healing energies in the center of the room. As the gold meets the green, you see it sparkle and gleam, and particles of color bounce off of each other. But they do not blend their colors, they stay distinctive -- golden and green...

Begin to absorb some of the green light through your left leg, in through your toes...

Absorb some of the golden light in through your right toes and up your leg until the two colors come together in your abdomen. Feel that effervescence within yourself...

Then let that mingled light rise up through your body, and return to the center of the room through your arms and hands.

As the light circulates through you, you feel the warmth of the lovely gold, and the coolness of the sparkling green, and the balance of the two give you healing...

Then allow some of the combined light to enter through your brow, to fill your head, to flow down through your throat, and again out your arms. Allow that to circulate and to bring balance into the body...

Now that you have washed your interior clean, your thoughts, your emotions are pure, invite your loved ones to come and share in this healing energy. (Take time to speak names aloud)

Welcome all of those who choose to come. Even as these energies are shared, accept the blessing for yourselves...

For the community of desire for peace in the world, allow this created ball of healing energy to spread out, to move around the world as though it were a cloud, depositing itself in those various parts to which it is called...

Release to themselves those who have chosen to join you.

Each of you become aware again of your own individual auric egg -- its rhythm, its balance, its protectiveness.

In acknowledgment of the universal energy which enables you to be healed and to be healers, close with singing. Blessed be. Go in peace.
(Ref. S-38 Sheariam)

Finding One's Personal Gem

When you are ready, settle your bodies comfortably. Breathe deeply and relax... In the first part of your meditation, join me on that infinite road which we walked together toward the pyramids. But tarry a while with me on the road. Do not rush to the pyramids...

As you move along the road, look to your right, look to your left. Let your eyes search out a glimpse of brilliance. For somewhere beside that road lies a precious gem which is for you and you alone. You will be given instructions later how to use that gem. We will give you a few moments of silence now, to look, to find, and to claim....

When you have found your gem, hold it gently in your hend. Examine it closely with your eyes. Feel its weight.. Enjoy its brilliance. Try to become aware of the subtle vibrational emissions from the stone...

Those of you who have your stones, find a convient place within your clothing to put them away. Those who have not found their stones, continue with this part of the meditation on your own until you do. Work with your stones a few minutes each day, becoming acquainted with them, learning to synchronize your energies with those emitted by the stone.

When you have your stones carefully put away, return your consciousness to this room, to your current bodily status, and become aware of the widening pool of energy that is being created in the center of this circle by the healers from the Temple.

These healing energies are warm, soothing, and comforting. And as more and more is poured into the pool, it begins to lap gently at your toes, and then rises warmly and soothingly to your ankles. The healing energies move up through your body from the pool through your feet...

As your body is permeated with these healing energies, you become more relaxed, more gentle with yourself. As your physical body becomes more comfortably relaxed, so your mental body relaxes too. The emotional body relaxes as well. The desire body becomes involved with the need for purity and strength and wholeness...

Whatever is needed to create this wholeness within the body is drawn by the desire body out of the pool of healing energy. It is drawn out of the pool and into your body, and you are healed... You are healed physically, mentally, emotionally, and spiritually...

You cannot be whole without being joyous as well. You also know that you must share this joyousness and the opportunity for healing. So you call to those consciousnesses of your loved ones to join you here. (Take time to speak the names aloud.)

As those who have been called choose to come, the healers move behind you. You may feel the presence as they bless you with their touch....

Even now you are being joined by powers from other realms. Listen closely. Perhaps you will hear some of the music. A blessing of light is poured upon you. And now, like the sunlight, this healing blessing is being spread around the world...

Release your loved ones to themselves. Know that you remain part of this great effort that goes forward. Feel the uplifting. Do not be surprised if you have dream visitors following this meditation. Blessings and love. Go in peace.
(Ref. S-44 Sheariam)

Charging Own Gemstone

Relax your bodies. Visualize in the center of your midst a three-legged stand upon which sits a totally transparent piece of crystal in the shape of a sphere. Within that sphere are no impurities. It is of such purity that light passes through without refraction...

Now at the center of the sphere, observe a tiny glow, like an incandescence... See that tiny glow grow in volume and in power until the crystal sphere lights up and glows like a sun -- pure white light, no refraction...

Now, those of you who have stones, hold them in the light glowing from that crystal, and allow them to be charged. The light is so pure, there can be no negativity left in your stone when charged... Now put your stones safely away, and continue to contemplate the glowing sphere...

Feel that pure radiation bathing you inside and out, purifying your physical organism, purifying your desire body, purifying your emotional body, purifying your mental body, until you, too, radiate that same pure radiance....

Into this magnificent radiance invite those for whom you have concern..... As you continue to invite your loved ones, and they continue to join you, continue also to contemplate the radiating light, for as you focus your attention upon it, it expands, it grows and grows, outward until the atmosphere of your planet is also purified with its pure light.

Then it moves beyond the physical into the Astral Realm to bless those who look for the Corridor of Transition,... to neutralize those forces which wreak havoc upon your planet...

Moving through the bodies of creativity... through the Realms of Soul and Mind... until Spirit is once again fed with its own light...

In that connection, you are blessed; you are protected; you are guided. All of your source groups, all of us in the Temple, join in this effort of purification....

Now gently return the focus of your attention to the here and now. Clasp within yourselves that blessedness and healing. Release your loved ones to return to themselves... Go in peace. May your lives be filled with joy.
(Ref. S-78 Sheariam)

Healing With Your Gemstone

Relax your bodies and be comfortable. Breathe deeply. Cleanse and purify yourselves... Set aside your thoughts of daily concern. For a moment take out your gem stone. Feel it in your hand. Enjoy the depth of its color and beauty..

Now choose some particular bodily ailment that you wish to work on, and press your stone to that part of your body. Allow your gemstone to become the conductor, the focalizing agent of the healing energies soon to be brought to you.

As the stone touches your body, you will begin to feel a pulsation. The pulsation of your body accords itself to the rhythm of the stone...

Now the healers join the circle, and to each individual comes his own particular healer. The healer covers the stone with his own hands... the pulsation becomes a throbbing as the pulsation of your body, and the stone, and the healer come into accord with each other...

As these energies come into alignment, one with the other, the proper state of health is made known to the tissues of the body. Where the tissues have gone astray, they come to know what is truth for them. The energy of the healer calls forth that truth... Even as you understand how much more comfortable and peaceful it is to function in the right place in the right way, so all the cells are made to understand that the proper function is the most desirable and the most enjoyable...

For a few moments of silence, allow this instructional process to go forward... Now bless your body, thank your healer, and take your gem stone between the palms of your hands. Feel the pulsing energy. Allow that energy to move to the center of the room, forming the pool of the waters of healing... The healers join you, contributing their energies to the pool. There is great power now in the pool of healing for general reorganizaion.

There is healing there not only of the physical, but of the mental, the emotional; healing for relationships, as well as disabilities.

You are welcome to invite those about whom you have concern to wade into this pool of reorganizational energy... Do not forget the members of your group who are absent... (Take time to speak names aloud.)

Put your own feet into the pool and take of the general healing... Feel the warming stimulation as the energy moves through your feet... up through your legs... up through your body... and from the core of your being emanates out into the other subtle bodies.... Enjoy the peace, the balance, and the harmony that is here available to you...

Carefully put your gemstones away. Release your loved ones to return to themselves. Crimp your toes, and touch your fingers together locking within your bodies the healing that you have taken from the pool...

Now with the ultimate love for your planet and its inhabitants, imagine the pool of energy being absorbed into the atmosphere, even as water is taken out of the ocean into the clouds. Make that image within your minds -- this healing energy being carried through the atmosphere completely around your globe, to be available to any who would bathe in its energy, to bring truth, and peace, and harmony to all the peoples of your planet.

Carry in your consciousnesses the thought of cooperative effort. Make of that a little mantra as you go through your days -- "Let there be cooperative effort". Go in peace.
(Ref. S-45 Sheariam)

Smoothing The Aura

(Meditation for an absent member who has asked for help.)

Settle your bodies. Prepare your minds. Imagine that you are bodiless, and thus able to move freely through the atmosphere of your planet... Imagine yourselves rising from your body, leaving that physical form seated where it is.

As you depart from that shell, you experience a great expansion — as though the bodies that you are have been confined within a very small area... Get used to the feeling of freedom, of expansion. Know that thought is your propellant — that you can think yourself anywhere you wish to be. Play with that for a moment or two...

Now gather round and together visit any absent member that has asked for your help. Approach quietly, softly, gently. Encircle your member. Join your hands as you do in your circle with your member in your midst, and pass energy from hand to hand as you have been taught to do*.

---

*(Join hands so that each has the right palm facing upward and the left palm facing downward. Relax the right hand, and squeeze the left hand.)

As the energy passes from one hand to the next and around your circle, it gains momentum and force...

Now release your hands and turn them palms forward toward your member in the center, releasing the energy that has passed through the circle into the member's subtle bodies to use. You have been asked to help, and now you are helping.

See the subtle bodies as a puffy mist around your member. With your hands, smooth the ragged edges. Pat and stroke... Stroke those subtle bodies as you would the arm of a beloved, soothingly, lovingly, healingly...

Now together say "I love you, (name of member); I love you, (name of member); I love you, (name of member)."

When you have finished, gently withdraw from the member, returning your awareness to your current placement. Bless yourselves and your efforts. Accept healing for yourselves. Go in peace.
(Ref. S-67 Judith)

Help For Unexpected Death

(Especially for sudden, explosive death)

Settle your bodies. Prepare your mind. Imagine you are bodiless -- able to move freely through the atmosphere of your planet..... Imagine yourselves rising from your body, leaving your physical forms seated where they are.

Go to that spot in your atmosphere above where the explosion took place. Join hands in a circle, and dance -- skip -- sing. Bring joy to that area. As you do this, the energies of the area become enlivened, and the traumatic shock which it has suffered is reduced. The parts of the subtle bodies of the departed which are ragged and torn are encouraged to join the other parts, to draw themselves back together.

In your dancing and your singing and your joy is great healing for them. It helps to offset all the thought waves of grief and longing and tragedy. For they not only suffer from their sudden demise, but their disappointment. Their goals and their accomplishment of them have been interrupted.

From your joyous singing and dancing energy they can draw support and healing, and get their thoughts back together so they can know what they must now do... (Pause long enough here to permit a good visualization and real participation.)

Now like children for whom the recess bell has rung, you must wave goodbye. Wish them well. Tell them that you love them. Say out loud, "I love all of you ".

Bring that joyous energy back to this room. Continue the mood of singing and dancing and joy. Join your hands in your imagination. Dance and sing and be joyful in your circle. You are joined by the healers and members of the Temple. Everyone dances and sings and is joyful with you. There is great healing here for all. You are blessed.

Together say,"I love"... Again. "I love." Again. "I love".... And now it is time to return your consciousness to the confines of your body. Keep within yourselves the joy and the healing, that you can treasure it in the days to come. Bless you.
(Ref. S-67 Judith)

Death And Transition

(Joy & love to a departed one and the family)

Relax your bodies and breathe deeply several times to oxygenate the brain tissue and to relax the abdomen... To the greatest extent possible, free your minds from your bodies, and feel that part of yourself which you call "spirit" rise and expand and experience freedom...

Feel yourselves lighter than air. Know that in this portion of yourselves lies the capacity to move to any part of your planet that you so direct it by your thoughts. Go first to any member of the family, and there surround that person with yourselves. Join your hands as you are wont to do... Be joyous, be gay, be loving, and be light-hearted...

Although you might not be able to picture where the physical body may be, envision the family member in your midst. Dance about and sing "(Name) I love you". (three times) (Repeat this for all the family members that you know).

In your minds, bless the bonds this family has established between its members, and reassure the departed member that those bonds shall continue even as he (or she) goes about the obligations which must be fulfilled...

With the love that you have extended to the family, the departed one comes to know that the family is supported, that he (or she) is not alone, and that the family is not alone...

Both the family and the departed one come to know that any of them are welcome to call upon your psychic support, your loving heart, your companionship...

The Temple doctors go to serve the one in transition. As you feel the freedom of your bodies, celebrate the freedom that that one now has from any pain-filled condition. Know it can now be released and it is all right. The departed one may now move toward joy.

Because he (or she) is a part of us now, we all, in our subtle bodies, in our hearts and minds, go forward with him (or her) until he (or she) enters that corridor of transition. This entry cannot be made until the passage rites are fulfilled.

Until that time, continue to send your love, your compassion, your mercy. Above all, as you think of the departed one, or as you think of the family members, hold only joy in your thoughts. Let the departed one be lifted by your thoughts of joyousness...

Now we invite you to bring your circle back to this room. Remain outside the prisons of your body. Remain free and joyous, and invite your loved ones to come and dance and sing with you...

Come, come kick up your heels. Be joyous....

The Temple musicians join you in your closing song as you return to your daily consciousness, and your loved ones return to theirs. Bless you. (Ref. S-68 Judith)

GLOSSARY

Absolute, The A "no-thing", eternal, infinite, unmeasurable, indescribable, imperishable, (S-19); that all-encompassing force or energy source that we call God, but much more than the conscious physical mind can comprehend.

Alleluia A song, or hymn, often sung by our circle, especially in closing or at times of stress.

aum A simulation of the sound of the universe; the Divine Roar; by "sound" definition, "The Inner Fire, which is the reflection of the Divine Fire, unites with the God sound, the roar of creation." (S-25). Pronounced "Ah-oo-mm".

Auphonim One of the Orders of Angels in the hierarchy given with the Laws; the name means "the wheels", and is associated with the Law of Ideation, the Principle of Migration, and the Attribute of Wisdom.

aura The "surround" of presence, color, energy and/or force which occupies the space immediately around a physical organism.

auric egg The external limit of the Vital Body as it meets with the Creative Body.

ball of energy By following the directions given by the teachers, one can produce a spherical force field between the palms, which then can be manipulated in various ways.

band Each individual has a group of "spirit" companions, a Source Group, sometimes referred to as one's "band".

bless While we cannot do anything for another, we can help them do for themselves by sending them our blessing, our mental wish for their wholeness and well-being, our appreciation of them.

chakra Vortices of sensitivity to various influences, energies, etc. There are seven (Crown, Brow, Throat, Heart, Solar Plexus, Spleen, Root) related to various activities and powers of the human body.

channeling The act in which the personal awareness recedes, permitting another consciousness to communicate using the brain and body of the individual. In this instance, the Temple teachers channel through Gloria Ruth.

Components, The Divine Germ Cell See Divine Germ Cell

Consciousness One of nine components of The Divine Germ Cell; the capacity to draw together the attributes developed through living.

Cosmic; Cosmos In the structure of existence the Realm of Cosmos is the first order, coming out the non-organization or chaos of the Realm of Divine which is the first emanation of The Absolute.

Disciplined Forcefulness The Attribute to be developed under the Law of I Am.

Divine Dewdrop The first emanation out of The Absolute; the first "thing" out of the "no-thing"; "prime mover of the universe" (S-19) comprising the Realm of Divine.

Divine Fire The enlivening force or energy which sustains Life.

Divine Flame See above

Divine Germ Cell, The At the instant of creation came forth from The Absolute the Divine Dew Drop emanating at Omega Points everywhere, all the time, comprised of nine Components which are ever-present in every thing. These nine components are: Two Opposing Forces, Two Minds (Perceiving and Creating), Life, Quality, Quantity, Volition and Consciousness.

Divine Will "The plan established by The Absolute for each person whether in flesh body or spirit" (S-32); later we understood that such a plan exists for every "thing".

energy Energy is referred to in many different ways throughout the teachings. Mainly, per Silverthistle, it is that quantity of Life (The Divine Germ Cell Component) that is granted to each creation or life form.

Blue and green healing energy is the substance of Life reduced to our dimension, channeled through our bodies, and seen as blue and green lights, sometimes other colors.

Eternal Life The true reality. Life itself is non-extinguishable.

Flame of Eternity A poetic reference to the Divine Fire

fontanel In an infant, there is a soft, pulsating, unosified area in the skull. This area closes and the bone hardens in the adult. The meditation refers to this area of the skull where the infant's "soft spot" is found.

Forces, The Two Opposing Components in the Divine Germ Cell.

Gabriel The second Archangel in the hierarchy of power given by Judith; a power center of strength.

God Self The intrinsic Divine Germ Cell essence of the Self.

guided meditation Meditation led by one individual who either reads or recites the meditation; the form used by the Temple teachers to keep every participant highly focused.

guides Non-earth-physical entities who are specifically committed to the welfare of a single human being.

Hallelujah Another spelling for Alleluia.

healers Entities of The Temple and/or members of a Source Group whose specific interest and duty is to assist in the establishment and maintenance of wholeness.

Higher Self A portion of the Self which has a wider perspective, clearer understanding of the individual's purpose in life.

homily A short dissertation.

initiate One who has studied a certain discipline and is actively engaged in applying it in his life. Initiates in the Temple of Radiant Life are living the radiant life. We who are students of this teaching are novitiates in the Temple.

Invisibles Beings who live in realms beyond the organic-physical spectrum.

invocation The act of calling upon powers from a higher source. In invoking the archangles, it is to activate power centers within ourselves.

Khamael The first Archangel in the hierarchy of power as given by Judith. The name means "heat at the core" or the Shadow of the Divine Fire manifest in life.

kundalini A Sanskrit word for the reservoir of life force which rests at the base of the spinal column.

Law of Harmony Sixth Law of the hierarchy, fourth law of our Universe; Principle is Revelation; Attribute is Knowledge.

Life, The Divine Germ Cell Component

Malachim One of the orders of angels, associated with the Law of Process, the Principle of Differentiation and the Attribute of Beauty. Subject to the Archangel Gabriel. The name means "royalty, or the reigning one"

mantra A word which has meaning by virtue of the sounds of which it is comprised; the sound of the mantra brings changes in consciousness or awareness.

Mind, Realm of One of the realms of Being and Creation operative in our universe.

Mind, Perceiving and Creative Two components of The Divine Germ Cell.

monitors Another name of one or more of one's Source Group.

novitiate A student on the pathway of learning that leads to the status of Initiate.

Orders of Angels Beings who exist in the Realm of Spirit, who do not incarnate, who carry out the missions designated by the power centers (both the Archangels and those in creation to which they are connected).

plutonian A devised word for the powerful "dark" side of earthly nature.

Principle of Differentiation The Principle upon which the Law of Process operates to develop the Attribute of Beauty; the operative for the Law of Process.

Principle of Migration The Principle upon which the Law of Ideation operates to develop the Attribute of Wisdom; the operative for the Law of Ideation.

progenitor Forebear; forefather; creator from the Realm of Spirit.

Quality A component of the Divine Germ Cell. Quality manifests as "tree-ness" or "human-ness" or "rock-ness" which is given a certain amount of "energy" or life force.

Quantity The component of the Divine Germ Cell which determines what form or "Quality" a creation will take.

quantification The increasing of one's "quantity" so as to further advance "Quality".

Realms Spheres of Being and Creation which stem from The Absolute. We have been taught about nine by the Temple teachers.

Relinquishment, Law of The ninth law having to do with Self-mastery which is what the teachers tell us we are supposed to be striving toward.

Self The entity-ego of which we as individuals are one personality.

self The singular personality manifest in this earthly existence.

spirit-physical A state of being in which a part of the entity is still physical but not visible to most, and a part is advanced to a greater understanding and wider perspective.

Spiritual Guide Same as member of a Source Group; one of the "band".

subtle bodies From the core of our being and extending outward beyond the boundaries of the known physical body are substantial bodies, physical in nature, usually unseen except by sensitives. Immediately next to the known physical is the Desire Body. Next outwardly is the Emotional Body; then the Vital Body. These are all peculiar to an individual. Beyond the Vital Body exists a Creative Body and a Mental Body which are shared by all creatures of this Realm of Self.

Temple of Initiation A singular organization of teachers, healers and others who work with less evolved realms to assist the inhabitants there to "quantify their Quality" and consequently further evolution.

trance channel The person who channels. See "channeling".

trance medium Same as above.

transition Usually called "death" by those who live on this planet.

transmutation The changing of one thing into another.

Tzadkiel One of the Archangels; the name means "to intuit".

Universe, Realm of One of the realms of Being and Creation created by Cosmos and creating out of itself the realms which make up our existence.

universe A general term usually meaning that which everyone and everything experiences in common.

Universal That which is of the Realm of Universe and manifesting in this Realm of Self.

universal The adjective of the general term above described.

Volition A component of the Divine Germ Cell.

APPENDIX

From time to time the Teachers have given us instruction on how to do or create certain things. There being no other seemingly "best" place for this information, we give it to you here.

ENERGY

In the meditation on Page 3 is given "Work with the ball of energy between your palms..."

You may begin to build this ball of energy by rubbing your palms together and then letting them separate gradually as the energy builds between them. See meditation on Page 47 for a direct quotation from the teacher Sheariam.

Another method is to circle your hands slowly, as though rubbing them together but with palms open and facing each other about an inch or two apart, until the energy builds and begins to feel like a ball between your hands. As you circle your hands, you will also feel a pulsing sensation. This is often quite strong and invigorating.

PYRAMIDS

On Page 65 you will find a meditation which describes how to build a golden pyramid around your head. With a little imagination, you can build this pyramid elsewhere as well for some amazing experiences.

INVITING LOVED ONES

In nearly every meditation we are told to include our loved ones, or those for whom we have concern. We have been taught that we cannot help another until requested to do so. Neither can we invoke for another, nor heal another. But we are free to invite anyone to join us in our meditations and healing circles by simply saying their names aloud. Then, we are told, those who wish to accept our invitation will do so and come to the circle in their consciousness to benefit in whatever way they desire.

TEMPLE HEALERS

We are often told that temple healers are joining us in our healing circles, and many of us feel their touch. You, too, can have this assistance in using these meditations. Sheariam has told us that all we need do is ask for such loving assistance and it is given freely, even when we are not in the circle and/or a Temple teacher is not present.

ASKING

If you want the attention of anyone from the Temple, simply address the issue in your thoughts, perhaps as follows:

"My body is in pain. I call upon the healers of the Temple of Initiation into The Radiant Life for help." Then quietly await their presence. Be accepting and appreciative.

You may call upon the teachers in the same way if something needs clarifying, or you feel you need some guidance.

When you are ready to conclude your meditation and turn your thoughts elsewhere, express your gratitude for their help. Then go on about your business. You will be surprised at the results.

INVOKING

As Judith and Sheariam gave us the information about the Nine Laws and the Archangels, they also gave us the meaning of sounds. In a number of these meditations you will find references to invoking, angels, archangels, etc. It is not within the scope of this work to give a comprehensive overview of these items. However, in the glossary is given the pronunciations for the few words and names used in this collection.

THE AMETHYST LIGHT

This meditation was given to the circle right after the Chernobyl "accident". During the preparation of this manuscript, Gloria Ruth was assured by the teachers that it could be inicluded for anyone to use, as the Temple's goal is to help any and all who would wish to better themselves. One of the circle members was also advised to hold an amethyst crystal in his hand while he was having X-rays taken, as amethyst is a specific against radiation damage, whether it is in crystal form, colors worn, or light.

(Note to editorial committee: Can you think of anything else that should be covered?)

INDEX

(Numbers refer to the first page of the meditation)

NOTES

COSMIC CONNECTION

A quarterly newsletter of the times
dedicated to presenting
highlights from the teachings.

Our staff believes that by giving we shall receive, thus enabling us to give more and more. Help us; then we may both help others.

SUBSCRIPTIONS

( ) Regular	$15.00/1-yr	$25.00/2-yr
( ) Sponsoring	$25.00/1-yr	$40.00/2-yr

A **Sponsoring Subscriber** receives a copy of each and every issue of Cosmic Connection, information on all events, study groups and open circles, and **discounts** on seminars and workshops.

To Purchasers of this limited edition book only we are offering new **Sponsoring Subscribers** the opportunity to purchase a complete set of the unabridged transcripts.

Circle Transcripts #2 to #92 $130.00
(Individual past transcripts are $2.00/ea. That's double your $25.00 back and you get four issues of Cosmic Connection as well!)

A **Sponsoring Subscriber** who purchases the complete set of transcripts may also subscribe to the regular circle transcripts as they become available (usually twice a month) for $35.00 paid in advance (including postage and handling). Twenty issues guaranteed or your unused balance refunded.

NAME____________________________

ADDRESS__________________________

CITY____________ST____ZIP_________